Global Issues in Water Policy

Volume 18

More information about this series at http://www.springer.com/series/8877

Min Jiang

Towards Tradable Water Rights

Water Law and Policy Reform in China

Springer

Min Jiang
College of Business
Victoria University
Melbourne, VIC, Australia

ISSN 2211-0631 ISSN 2211-0658 (electronic)
Global Issues in Water Policy
ISBN 978-3-319-67085-0 ISBN 978 3 319 67087-4 (eBook)
DOI 10.1007/978-3-319-67087-4

Library of Congress Control Number: 2017952349

Printed on acid-free paper

This Springer imprint is published by Springer Nature
The registered company is Springer International Publishing AG
The registered company address is: Gewerbestrasse 11, 6330 Cham, Switzerland

Preface

Coming from a water abundant city in South China, I did not realise how water presents an immense challenge for China until I had lived in the extremely water scarce capital city, Beijing, for my postgraduate study at Tsinghua Law School. The first spring that I spent in Beijing remarked my first life experience in yellow dusty sandstorms, which later I learnt were mainly caused by aggravating desertification as a result of overexploitation of water resources in China's second largest inland river basin, the Hei River Basin. In 2005, when I had the opportunity to undertake a PhD in environmental law at Macquarie University in Australia, I was inspired to focus my study on water law and policy reform.

Freshwater resources in many countries around the world are increasingly pressured by population growth, economic development, urbanisation, rising standards of living, changing values of the environment, and on top of all these stresses, climate change. While governments confront major issues in sustainable water use, how water governance responds to water scarcity, variability and uncertainty has become a key global policy challenge. Water reforms have been observed globally, delivering mixed outcomes. The roles of government and market in water allocation lie at the centre of the water reform debate, reflecting different values of water as both a public/social and economic good. China's current water reform aims to explore the application of market-based water allocation mechanisms, for which it is recognised that learning from international experiences is important. Against this backdrop, my PhD project provided a first comprehensive legal examination of water rights arrangements and water rights trading in China by reflecting on experiences from overseas jurisdictions, and particularly Australia and the United States. My PhD thesis forms the foundation of this book.

My interest in water governance continues into my post-doctoral research. As water reforms in both Australia and China deepen, I see the need for an up-to-date reflection on the recent progress in the development of water markets in the two countries. In 2015, as the chief investigator, I led a research project which focused on what and how China can benefit from the Australian experiences to address its water governance challenges in developing functioning water markets. This book is also informed by the research findings of this project.

The central question that the book attempts to explore is how law and policy can facilitate China's transformation in its approach to water allocation from an extreme government-based mechanism towards increased application of market mechanisms. Based on in-depth reviews of China's current legal and policy frameworks for water rights arrangements and trading comparative to international experiences, this book proposes potential solutions for China's water law and policy reform to move towards tradable water rights. It is hoped that the book will add important components to the existing body of knowledge in water rights and trading by reflecting on not only China's water reform but also the international trends in water governance. To this end, this book is for readers who are interested in the growing theoretical and policy challenges around water governance in general, and development of water markets in particular. Although the book takes a legal and policy perspective, it may be of interest to readers across a range of disciplines, including environmental (water) law and policy, environmental studies, water resources management, public policy, and Asian (Chinese) studies. China has become the centre of the world's attention as its actions in sustainable development have significant implications for the rest of the world. It is expected that the insightful examination of water law and policy reform in China in this book will help international audiences to better understand China in transition.

I would like to acknowledge the financial support that helped me undertake research related to this book. My PhD project was funded by an International Postgraduate Research Scholarship (IPRS) and International Macquarie University Research Scholarship (iMURS). In addition, an International Collaborative Research Grant between Victoria University (VU) (Australia) and Central University of Finance and Economics (CUFE) (China) supported me to lead a joint research project entitled "Developing functioning water markets for sustainable water use: Australian experiences and implications for China's water reform".

I am indebted to my PhD supervisors, Professor Michael Jeffery QC and Professor Donna Graig, for their valuable guidance and support during my study at the Centre for Environmental Law, Macquarie University. I am also truly appreciative of the contributions that my colleagues have made to the VU-CUFE joint research project. At the VU side, Professor Elisabeth Wilson-Evered and Dr. Shah Miah have brought their expertise in leadership and information systems to the project, and Ms Li He has contributed to the project management and data analysis as a research assistant. At the CUFE side, Associate Professor Qian Liu and Dr. Yinshuo Xu have played an instrumental role in interview data collection and organisation of the focus group. My gratitude extends to those who participated in my PhD fieldwork as well as the VU-CUFE project. For confidentiality reasons, I am unable to list their names here, but their contributions are greatly appreciated.

I would like to acknowledge assistance that I received from Fritz Schmuhl, Senior Editor, and Joseph Daniel, Project Co-ordinator from Springer. They have been extremely supportive and understanding, and assisted me to eventually deliver this book after an extended delay due to my maternity break. I would also like to thank Ms. Jeanne Enslin for her superb proofreading service for the book.

I dedicate this book to my family. My 21-month-old son, Brian, brings me endless joy every day. Snuggle time before bed at night was the most recharging break off my book writing. My husband, David, has fully supported me by sharing many baby duties and freeing me from chores. My parents accompanied me in the last and the most difficult months of my PhD study. This book is a result of the lasting commitment, patience, and encouragement of my family. It would not have been possible without their unconditional support and sacrifice.

Melbourne, Australia
May 2017

Min Jiang

Contents

Abbreviations

ACT	Australian Capital Territory
BoR	Bureau of Reclamation (California)
BYRPM	Bureau of Yellow River Project Management (Inner Mongolia)
CDWA	Competent Department of Water Administration (China)
CESCR	Committee on Economic, Social and Cultural Rights (UN)
CEWH	Commonwealth Environmental Water Holders (Australia)
CLAPV	Centre for Legal Assistance to Pollution Victims (China)
COAG	Council of Australian Governments
CSOs	Civil Society Organisations
CWCB	Colorado Water Conservation Board
DGA	General Water Directorate (*Dirección general de Aguas*) (Chile)
DPI	Department of Primary Industries (New South Wales)
EPB	Environment Protection Bureau (China)
FAO	Food and Agriculture Organisation (UN)
FYP	Five Year Plan (China)
GDP	Gross domestic product
GWP	Global Water Partnership
ICESCR	International Covenant on Economic, Social and Cultural Rights
IDSB	Irrigation District of the Southern Bank of Yellow River (China)
IMDWR	Inner Mongolia Department of Water Resources
IMEPC	Inner Mongolia Electric Power Corporation
IPCC	Intergovernmental Panel on Climate Change
IUCN	International Union for Conservation of Nature
IWRM	Integrated Water Resources Management
MDB	Murray Darling Basin (Australia)
MDBC	Murray Darling Basin Commission (Australia)
MDGs	Millennium Development Goals
MEP	Ministry of Environmental Protection (China)
MLR	Ministry of Land and Resources (China)
MOF	Ministry of Finance (China)
MOHURD	Ministry of Housing and Urban-Rural Development (China)

MWR	Ministry of Water Resources (China)
NBS	National Bureau of Statistics (China)
NDRC	National Development and Reform Commission (China)
NFA	National Forestry Administration (China)
NGOs	Non-Governmental Organisations
NHFPC	National Health and Family Planning Commission (China)
NPC	National People's Congress (China)
NSW	New South Wales (Australia)
NTA	Native Title Act (Australia)
NWC	National Water Commission (Australia)
NWI	National Water Initiative (Australia)
OBIM	Ordos Bureau of Irrigation Management of the Southern Bank of Yellow River (Inner Mongolia)
OECD	Organisation for Economic Co-operation and Development
RICDs	Recreational in-channel diversions
SNWDP	South-to-North Water Diversion Project (China)
SWRCB	State Water Resources Control Board (California)
UNDP	United Nations Development Programme
UNEP	United Nations Environment Programme
US	United States (of America)
WAB	Water Affairs Bureau (China)
WRB	Water Resources Bureau (China)
WUA	Water Users' Association
WWAP	World Water Assessment Programme
YRCC	Yellow River Conservancy Commission (China)

Chapter 1
Introduction

Abstract This chapter is an introduction to the entire book. It starts by discussing the state of the global water crisis which raises challenges to enhance water governance from both international and Chinese perspectives. It briefly explains why the book is important, what it aims to contribute to the existing scholarship, and how the content was researched. The chapter also provides an overview of the book.

Keywords Water crisis • Water governance debate • Water reform • Water allocation mechanisms • Water markets

1.1 Problem Statement

1.1.1 The International Perspective

As an essential resource for life, water has often been taken for granted, as if there were an abundance of water available for every claim of use. It is true that two thirds of the earth's surface are covered by water, but most of this is not usually referred to as a water resource.[1] Accounting for 97.5% of the total volume of the world's water, the water in oceans and seas is too salty for most productive uses.[2] Approximately 69% of the remainder is locked up in ice sheets and glaciers, and considered inaccessible and not available for use.[3] Another 30% is groundwater.[4] Holding only 0.3% of the earth's total freshwater resources, surface waters including lakes, rivers, streams, wetlands and reservoirs, represent about 80% of the annually renewable

[1] World Water Assessment Programme (WWAP), *Water for People, Water for Life,* the United Nations World Water Development Report 2003, Paris: UNESCO and Berghahn Books, http://www.unesco.org/new/en/natural-sciences/environment/water/wwap/wwdr/ (Accessed November 17, 2014), 76.

[2] Ibid.

[3] WWAP, *Water, a Shared Responsibility*, the United Nations World Water Development Report 2, 2006, Paris: UNESCO, http://www.unesco.org/new/en/natural-sciences/environment/water/wwap/wwdr/ (Accessed November 17, 2014), 121.

[4] Ibid.

M. Jiang, *Towards Tradable Water Rights*, Global Issues in Water Policy 18,
DOI 10.1007/978-3-319-67087-4_1

water.[5] Therefore, the availability of fresh water is much less than what one would have thought. Freshwater resources in many countries around the world are heavily pressured by population growth, economic development, urbanisation, rising standards of living, changing values of the environment, and on top of all these stresses, climate change. Water shortage, pollution, and deteriorating freshwater ecosystems are frequently reported around the world so that many believe that "the world is in a water crisis".[6]

While the existence of a worldwide water crisis remains questionable, it has been widely recognised by both research and policy communities, that the water crisis is essentially a crisis of water governance.[7] Water governance here refers to "the range of political, social, economic and administrative systems that are in place, which directly or indirectly affect the use, development and management of water resources and the delivery of water services at different levels of society".[8] Shaped and complicated by the mobile and fugitive nature of water, water governance is required to deal with water's variability and uncertainty, to balance multiple and diverse values of uses, and to avoid and/or solve conflicts that are more likely to intensify among increasingly competing water users. The recent manifestation of climate change creates new challenges for sustainable development of water resources as well as water governance. Climate change impacts on hydrological resources and patterns include alternations in the timing, spatial distribution, quality and quantity of water resources.[9] Moreover, climate change has complex interactions with other drivers of change through for example food and energy production and consumption.[10] In the context of increasingly uncertain future conditions of climate and water resources, how water governance responds to increasing water scarcity, variability and uncertainty is therefore a key global policy challenge. Water reforms have been observed globally that introduce an interesting mix of success stories and failures along with a common desire to achieve balance between increasingly competing values of water use.

Water governance is faced with a double challenge as water is both a social and an economic good. On the one hand, water is an indispensable social resource upon which humans rely in order to survive. On the other hand, with water becoming scarcer, its economic value has gained wider recognition in the international forums. Water's social and economic values are reflected in water governance debates on treating water as a human right versus allocating and pricing water as an economic good. Water as a human right requires affordable access to safe drinking water and sanitation, which means that water may be supplied as a public and sometimes even free resource, and water users only need to pay a proportion of the costs of water

---

[5] Ibid, 125.

[6] WWAP, *Water for People, Water for Life*, 10.

[7] WWAP, *Water, a Shared Responsibility*, 47.

[8] Ibid.

[9] Intergovernmental Panel on Climate Change (IPCC), *Climate Change and Water*, IPCC Technical Report VI. Geneva: IPCC, 2008.

[10] Ibid.

supply thanks to public finance subsidies. If water is treated as an economic good, however, the principle of full cost recovery applies. These two approaches take different, sometimes even opposite directions towards the roles that government and market play in water governance. Traditionally, water has been provided, managed and allocated with substantial involvement of government as a public good. The economic dimension of water, by comparison, favours market force in water allocation. The institutional alternatives for water allocation represent a spectrum from complete government control at one extreme through a mixture of market and government allocation in the middle, to predominant market allocation at the other extreme.[11]

Historically, government allocation has been employed as a main mechanism in many countries based on the need for protection of public values of water, which has worked well in many cases. However, the incentive structures within government could potentially generate "government (public) failure".[12] Some most common "ills of the public sector"[13] include: rent seeking in both economic and socio-political forms,[14] incentives not related to performance and efficiency,[15] special-interest effect (politically powerful interests and their clients dominating the policy arena and receiving political favours),[16] and short sightedness effect.[17] In response to addressing these government failures, market allocation mechanisms have drawn the attention of both researchers and policy makers.[18] The effectiveness of market allocation to direct water to economically higher valued uses has been supported by empirical evidence from a number of jurisdictions (such as Australia and western United States). Water markets and tradable water rights are two different yet overlapping concepts related to market allocation mechanisms. Water markets can exist in an informal fashion in the absence of a formal definition of water rights. For example, operating at the local level among tube well owners and other farmers, informal water markets in Pakistan and Bangladesh helped irrigators meet their needs in water shortage.[19] Central to the development of formal water markets,

---

[11] Ariel Dinar et al., *Water Allocation Mechanisms: Principles and Examples*, World Bank Policy Research Working Paper, No. 1779, 1997, 3.

[12] Terry L. Anderson, "The Water Crisis and the New Resources Economics," In *Water Rights: Scarce Resource Allocation, Bureaucracy, and the Environment*, ed. Terry L. Anderson (San Francisco, California; Cambridge: Pacific Institute for Public Policy Research; Mass, Ballinger Pub. Co., 1983), 5.

[13] C. J. Perry, Michael Rock, and D. Seckler, *Water as an Economic Good: A Solution, or a Problem?* International Irrigation Management Institute Research Paper, 14, 1997, 7.

[14] Dinar et al., *Water Allocation Mechanisms*, 7.

[15] Ibid. Also see Anderson, "The Water Crisis and the New Resources Economics," 5–6.

[16] Anderson, "The Water Crisis and the New Resources Economics," 5–6.

[17] Ibid, 5.

[18] K. William Easter and Qiuqiong Huang, eds. *Water Markets for the 21st Century: What Have We Learned?* (Dordrecht: Springer, 2014), Preface.

[19] Syed Z. Sadeque, "Nature's Bounty or Scarce Commodity: Competition and Consensus over Groundwater Use in Rural Bangladesh." In *Negotiating Water Rights*, ed. Bryan R. Bruns and Ruth Meinzen-Dick (London: ITDG Publishing, 2000), 269–291.

tradable water rights require "both the formal definition of entitlements, and the specification of the conditions under which the entitlement may be traded".[20] Water trading practices in Australia, Chile, and western United States (US) are prominent examples of formal and highly organised water markets.

In the international water debate, discussions remain continuously heated around government versus market as different approaches to water governance. The centre of the water governance debate lies in the need for a balanced combination of market and government water allocation institutions and mechanisms. A well-functioning water market requires government to provide underpinning and facilitating legal and regulatory infrastructure. In particular, tradable water rights need to be well defined and specified, secure and enforceable; the market needs to be regulated properly with a range of rules in place; buyers and sellers need to have easy to use trading platforms to trade, in which intermediaries may play an important role to facilitate transactions; and more broadly, the context of institutional and governance arrangements is crucial to remove barriers to water trading, to reduce transaction costs, and to encourage water users to participate in water trades. Across all these key elements of water markets, to address externality issues and risk of market failures, however, market allocation for water needs a higher level of government intervention than other economic goods. Water rights trading practices across jurisdictions have taken various approaches to government intervention in water allocation aiming to suit their specific local contexts. One thing is sure – no water market is perfect, with challenges remaining around reduction of trading barriers, management of externalities, and protection of third parties' interests. As a result, water reforms around the world are ongoing, as a process of exploring a balanced mix of water allocation mechanisms suitable for a specific local context as the level of water scarcity varies.

1.1.2 The Chinese Perspective

Leading the world's economic growth, China's development has seriously overstretched its natural resources. It is not surprising to see that China is facing one of the world's most severe water crises.[21] Despite its world leading rank in total water availability (ranked 5th),[22] China is dangerously short of water with its annual per capita water availability just one third of the world average.[23] Nine out of 31 provin-

[20] Perry et al., *Water as an Economic Good: A Solution, or a Problem?* 7.

[21] 2012 China State Council Opinions on the Implementation of the Strictest Water Resources Management System, State Council Document No.3, 2012, http://www.gov.cn/zwgk/2012-02/16/content_2067664.htm (accessed October 24, 2016).

[22] World Bank, *World Development Indicators: Table 3.5: Freshwater*, 2014, http://wdi.worldbank.org/table/3.5 (accessed October 24, 2016).

[23] Ibid. For details about China's water availability, please see Chap. 2, Sect. 2.1.

cial level regions face an "absolute scarcity" with per capita water resources well below 500 cubic metres by the international definition.[24] China's water scarcity challenges also consist of water quality issues. The water environment and ecosystems suffer from overexploitation and pollution, leading to worsened water shortage and public health problems. Climate change further exacerbates the pressure on China's water resources. Already, floods are observed more frequently and at bigger scales in China, consistent with the projected increase in annual average precipitation. However, it has been observed that the runoff in the Chinese major rivers has decreased in the last 40 years. Projections show that the amount of runoff is likely to decrease in the already arid northern China and increase in relatively water abundant southern China, which may intensify the water shortage in northern China.[25]

Exacerbated by climate change impacts, China's water crisis involves water shortage, pollution, and ecosystem deterioration, each presenting distinctive challenges for policy reform. China's water governance is therefore increasingly challenged to balance multiple and diverse values of water use. Dominated by the philosophy of "man can conquer nature", China's water governance was traditionally oriented towards construction and utilisation of water resources, instead of conservation and protection.[26] This traditional approach to water allocation and management failed to encourage water saving and water use efficiency due to its ineffective administrative allocation mechanisms and fragmented management.[27] In response to the worsening water crisis, China has initiated a transformation in its approach to water management from project-oriented towards resource-oriented, implying changes in its philosophy of water use. The relationship between man and nature is no longer considered absolutely against each other. Rather, resource-oriented water management calls for "harmonious coexistence between man and nature".[28]

It is within this context that the Chinese government has initiated an ambitious water reform, "Building a water saving society". This water reform aims to integrate the traditionally fragmented water management systems, establish a national water rights system, and implement market-based allocation mechanisms for more sustainable water use. Pilot programmes were launched and implemented to experiment

[24] WWAP, *Managing Water under Uncertainty and Risk*, the United Nations World Water Development Report 4, 2012, http://www.unesco.org/new/en/natural-sciences/environment/water/wwap/wwdr/wwdr4-2012/ (accessed October 24, 2016), 122.

[25] Rex Victor Cruz et al., "Asia," In *Climate Change 2007: Impacts, Adaptation and Vulnerability. Contribution of Working Group II to the Fourth Assessment Report of the Intergovernmental Panel on Climate Change*, ed. Martin Parry et al. (Cambridge, UK: Cambridge University Press, 2007), 469–506.

[26] Shucheng Wang, *Resource-oriented Water Management: Towards Harmonious Coexistence between Man and Nature*, 2nd ed. (Singapore: World Scientific Publishing; Beijing: WaterPower Press, 2006), 37.

[27] Min Jiang, "Water Rights Trading in China's Water Policy Reform: Current Practices and Future Development," *Proceedings of Water Down Under 2008*, Adelaide: Engineers Australia, 2008, 1831–1842.

[28] Wang, *Resource-oriented Water Management*, 37.

with water rights initialisation and trading, and new laws and policies were developed to provide regulatory frameworks for water markets. After nearly 15 years of major efforts in legal, policy, and institutional developments and piloting practices, China has established a preliminary framework for its national water rights system. Nevertheless, the development of water markets in China is still in its infancy. Water use security and water quality control remain immense challenges for China. In early 2012, the Chinese Central Government pushed the water reform forward towards more strict water governance, with the intention to refine its institutional arrangements of total quantity control of water consumption and quota management.[29] This implies that where the total quantity of water consumption and abstraction has reached or exceeded the capped total limit, no further water abstraction permits should be issued or approved. Water rights trading then becomes an alternative vehicle to obtain water use rights. China's water reform therefore calls for further development of market-based water allocation mechanisms, refining of the water rights system, establishment of water markets, and encouragement and facilitation of water trading.[30] To do so, Chinese researchers and policy makers recognise the importance of learning from, and building on the international experiences in water rights trading policy and practice. Indeed, as water reforms keep deepening in many countries around the world, an up-to-date reflection will be useful to shine some light on the way forward.

1.2 Aim and Scope

1.2.1 Research Aim

Against the backdrop of China's water reform and the wider international debate in water governance, this book aims to provide an innovative approach to the complex issue of water governance by critically analysing the recent legal and policy developments in China towards tradable water rights. It examines the deficiencies of the current systems for water rights arrangements and trading; explores how China may learn from and build on the international trends in water rights trading practices (mainly Australia and the United States); and proposes legal and policy frameworks for defining and administering tradable water rights in China that underpin sustainable water use in the face of exacerbated water scarcity, variability, and uncertainty. All in all, the book proposes pragmatic strategies for China's water law and policy reform to move towards tradable water rights, which encompasses a comprehensive prescription from initialising and defining tradable water rights to administering water rights and trading.

[29] 2012 China State Council Opinions on the Implementation of the Strictest Water Resources Management System.

[30] Ibid.

By reflecting on the deepening water reforms in both China and other jurisdictions, the book aims to contribute to the international water governance debate by exploring from a legal and policy perspective, how China, comparative to other cases around the world, can find a balanced combination of water allocation mechanisms to address its water challenges. As China's starting point for water reform is an extreme government control over water resources towards market allocation, how water markets can work functionally with adequate but not overreaching government intervention within China's context will be an interesting and challenging enquiry. Well aware that the effectiveness of water markets and tradable water rights in sustainable water use remains to be a continuing debate,[31] this book accepts the direction towards tradable water rights a given reality of China's water reform. It does not necessarily base its argument on market as the only or better solution for water allocation. Rather, it takes a pragmatic approach to the discussion about the ongoing reform in China by working with the fact that China is now moving from an extreme government-based water allocation mechanism towards increased application of market mechanisms. As a result, the book intends to examine how law and policy can facilitate such transformation, and help China's water reform find the balance between market freedom and government intervention. By exploring inspirations from international experiences suitable for China's context, it is hoped that the observations and proposed implications for China's water reform will contribute to developing a better understanding of the way in which experiences in water markets can be shared from jurisdiction to jurisdiction.

1.2.2 Research Scope

Water resources in this book only refer to fresh water, which includes surface water (rivers, lakes, and reservoirs) and groundwater (aquifers). As the book focuses on the domestic legal and policy issues of China related to water resources, transboundary water bodies that involve China and other nations are also excluded.

The concept of water rights is interpreted from different perspectives in the literature. In a simplistic view, it refers to rights to use water. Some studies have used the term in a broad sense when researchers have discussed water rights as a human right without considering specific types of use. In contrast, others narrow the definition down against a range of categories of water use. There are many forms of water use, consumptive or non-consumptive, such as abstraction, navigation, fishery, and so on. In some contexts, pollutant discharge is considered a type of water use as well. To keep the book within a manageable scope, water rights trading in this book only refer to trading in water abstractive rights, which lie at the centre of the most water conflicts in China. Despite the focus, the term of water rights may be used to refer to various meanings throughout the book.

[31] Easter and Huang, eds. *Water Markets for the 21st Century*, 13.

From an international perspective, success stories have been observed in some overseas jurisdictions where water markets are increasingly contributing to offering efficiency and flexibility for water users. Among others, the western US have long applied the water rights trading market mechanism to help relieve competitions for relatively scarce water among various communities including farmers, cities, and environmentalists.[32] More recently, Australia has been recognised internationally as a notable water reform success story.[33] The different levels of the development stage and sophistication of the water markets provide an opportunity for China to learn from overseas.

One may argue that the comparative analysis could hardly help as those jurisdictions under observation belong to common law systems, opposed to China's civil law system. However, such learning exchange across different law systems is useful for two important reasons. Firstly, most jurisdictions, whether from common law or civil law, tend to establish statutory systems to manage water resources and water use. This is mainly because they face similar challenges in water governance of which a critical issue is how to efficiently and sustainably allocate limited water among competing users and uses. Water allocation mechanisms deal with multiple values of water and set the rules for water use, lying at the heart of water governance. Evolution in water rights and markets involves adjustments and movements between different institutional arrangements of water allocation, highlighting the wrestle between government and market. Secondly, diversity in water rights law is not limited to the difference between the common law system and the civil law system. Rather, shaped by water's physical states, water use patterns, and other localised contexts of social, economic, and legal frameworks, water rights regimes are substantially diverse across jurisdictions, even those within a same law system. The eastern and western states in the US for example have developed significantly distinct water rights law as a result of differences in history in response to the relative scarcity of local water resources.[34] What is relevant is that common key elements of water market can be observed across jurisdictions, and reflections on international experiences help China identify its own governance gaps. The next step is how to address these gaps by referring to international experiences to suit China's specific context, which should take the potential need for localisation and modification into consideration. Bearing this in mind, the book attempts to examine the suitability of introducing an overseas practice in China and making recommendations for modification if needed.

This book also involves empirical analysis on the current practices in water rights trading in China, based on two stages of fieldwork. The first stage of the fieldwork was conducted in 2006, focusing on the first pilot practices of water rights trading

[32] Terry L. Anderson et al., *Tapping Water Markets* (New York: Routledge, 2012), Preface.

[33] See Easter and Huang, eds. *Water Markets for the 21st Century*, 46. Also see Debra Townsend and Odette Adams, "Water Rights and Trading in Australia," *Australian Environmental Law Digest*, Sep 2016: 3–4. http://0-search.informit.com.au.library.vu.edu.au/documentSummary;dn=316367663218575;res=IELHSS (accessed May 3, 2017).

[34] Anderson et al., *Tapping Water Markets*, 26.

in China. Within the context of the water reform, three types of market-like water transfers took place in different areas of China, namely water trading between local governments (the Dongyang-Yiwu case), water supply contract transfers between irrigators (water notes transfer in Zhangye), and water rights transfer from agriculture to industry in the Yellow River Basin. A series of face-to-face semi-structured interviews were conducted in Zhangye (Gansu Province), Erdos (Inner Mongolia Autonomous Region), and Dongyang (Zhejiang Province) in November 2006. Interview participants included representatives from three major stakeholder groups in those piloting practices, namely farmers, staff of water users associations, and local water officials, in order to collect information related to water rights, water trading and administration. Since the functions of water trading in improving water allocation efficiency and reducing water overuse have already been validly argued by many previous studies, this stage of fieldwork did not focus on how these pilot practices demonstrated their effectiveness in water allocation although the merits of water rights trading did come across as a strong message through the data collected. Instead, it aimed to examine how the functionality of water rights trading was undermined by the existing legal and institutional system of water management in China, which originally worked for complete government control, and thereby lacks sufficient legal and policy infrastructure to underpin and facilitate water rights trading.

The second stage of the fieldwork was undertaken in 2015, to obtain a more up-to-date understanding of China's water governance framework after more than 15 years of substantial efforts in the reform. The 2015 fieldwork included semi-structured interviews conducted between September and October 2015, and a focus group discussion held on 27th October 2015 in Beijing, with Chinese water stakeholders across government, industry, and academia. The data collected from the second stage of fieldwork informed an updated understanding of the most recent progresses and remaining challenges in China's water reform around the development of water markets. The fieldwork followed the requirements of the human research ethics clearance through the author's universities. Interviewees and focus group discussion participants in the two stages of the fieldwork remained anonymous in the book.

1.3 Overview of the Book

The book consists of eight chapters. The Introduction (Chap. 1) starts with the state of the global water crisis which raises challenges to enhance water governance from the both international and Chinese perspectives. It briefly explains why the book is important, what it aims to contribute to the existing scholarship, and how it was researched.

Chapter 2 provides a review of the recent developments in water governance around the implementation of market-oriented allocation mechanisms. Following a brief review of China's water scarcity, the chapter sets the theoretical framework for the entire book by reviewing the international debates on water reform. It describes water's characteristics as a resource and the complexity of water use, which have significant implications for shaping water governance. The recent developments in water governance from an international perspective are also examined in this chapter. It covers a number of key concepts which have significant implications for the legal and institutional arrangements of tradable water rights, including interdependency of water use, water as a social and economic good, alternative property regimes as a water policy instrument, government versus market in water allocation, and last but not least, climate change adaptation and integrated water resources management.

Chapter 3 reviews the current legal and policy developments in water rights trading in China, including legal, policy, and regulatory frameworks related to the key issues in water management. It provides an insightful understanding of the legal and institutional context within which tradable water rights have to be developed and operated. In Chap. 4, the book reflects on the current practices in water rights trading, in the context of other alternative water allocation mechanisms in China. Assisted by empirical analysis, such reflection is important to show that while the merits of water rights trading have been demonstrated through the piloting practices, its effectiveness has also been impeded due to the absence of basic legal and policy infrastructure. More importantly, major governance gaps are identified in the current systems for sufficient facilitation of water markets. Chapters 3 and 4 together serve as a starting point for future legal and policy reform towards tradable water rights in China.

Chapters 5, 6 and 7 critically analyse the recent legal and policy developments in China towards tradable water rights including initialisation (Chap. 5), definition (Chap. 6), and administration of tradable water rights (Chap. 7), and administration of water rights trading and establishment of water markets (Chap. 7). The chapters examine the deficiencies in the current systems, explore how China may learn from and build on the international experiences (especially in Australia and the US), and propose legal and policy frameworks for defining and administering tradable water rights in China.

The concluding chapter (Chap. 8) draws together the proposals that are made throughout the book to feed an integrated framework for China's water law and policy reform to move towards tradable water rights. It puts China's water reform in the context of its wider societal transition, as well as the wider context of international water governance discussions and policy challenges in responding to future changes in water supply and demand. It also points out directions for future research.

References

Anderson, T. L. (1983). The water crisis and the new resources economics. In T. L. Anderson (Ed.), *Water rights: Scarce resource allocation, bureaucracy, and the environment*. San Francisco/ Cambridge, MA: Pacific Institute for Public Policy Research/Ballinger Pub..

Anderson, T. L., Scarborough, B., & Watson, L. R. (2012). *Tapping water markets*. New York: Routledge.

Cruz, R. V., Harasawa, H., Lai, M., & Shaohong, W. (2007). Asia. In P. Martin, O. F. Canziani, J. P. Palutikof, P. J. van der Linden, & C. E. Hanson (Ed.), *Climate change 2007: Impacts, adaptation and vulnerability. contribution of working group ii to the fourth assessment report of the intergovernmental panel on climate change* (pp. 469–506). Cambridge: Cambridge University Press.

Dinar, A., Rosegrant, M. W., & Meinzen-Dick, R. (1997). *Water allocation mechanisms: Principles and examples, Policy research working paper, No. 1779*. World Bank.

Easter, K. W., & Huang, Q. (Ed.). (2014). *Water markets for the 21st century: What have we learned?* Dordrecht: Springer.

Intergovernmental Panel on Climate Change (IPCC). (2008). *Climate change and water, IPCC Technical Report VI*. Geneva: IPCC.

Jiang, M. (1831–1842). Water rights trading in China's water policy reform: Current practices and future development. *Proceedings of water down under 2008*, Adelaide: Engineers Australia, 2008.

Perry, C. J., Rock, M., & Seckler, D. (1997). *Water as an economic good: A solution, or a problem? International irrigation management institute research paper, 14*. Colombo: International Irrigation Management Institute.

Sadeque, S. Z. (2000). Nature's bounty or scarce commodity: Competition and consensus over groundwater use in rural Bangladesh. In B. R. Bruns & R. Meinzen-Dick (Ed.), *Negotiating water rights* (pp. 269–291). London: ITDG Publishing.

Townsend, D., & Adams, O. (2016, September 3–4). Water rights and trading in Australia. *Australian environmental law digest*. http://0-search.informit.com.au.library.vu.edu.au/documentSummary;dn=316367663218575;res=IELHSS. Accessed 3 May 2017.

Wang, S. (2006). *Resource-oriented water management: Towards harmonious coexistence between man and nature* (2nd ed.). Singapore/Beijing: World Scientific Publishing/WaterPower Press.

World Bank. (2014). *World development indicators: Table 3.5: Freshwater*. http://wdi.worldbank.org/table/3.5. Accessed 24 Oct 2016.

World Water Assessment Programme (WWAP). (2012). *Managing Water under Uncertainty and Risk*. The United Nations World Water Development Report 4. http://www.unesco.org/new/en/natural-sciences/environment/water/wwap/wwdr/wwdr4-2012/. Accessed 24 Oct 2016.

WWAP. (2003). *Water for people, water for life*. The United Nations world water development report 2003. Paris: UNESCO/Berghahn Books. http://www.unesco.org/new/en/natural-sciences/environment/water/wwap/wwdr/. Accessed 17 Nov 2014.

WWAP. (2006). *Water, a shared responsibility*. The United Nations world water development report 2. Paris: UNESCO. http://www.unesco.org/new/en/natural-sciences/environment/water/wwap/wwdr/. Accessed 17 Nov 2014.

Legal and Policy Sources

2012 China State Council Opinions on the Implementation of the Strictest Water Resources Management System, State Council Document No.3, 2012., http://www.gov.cn/zwgk/2012-02/16/content_2067664.htm. Accessed 24 Oct 2016.

Chapter 2
Water Governance in a Changing World: China and Beyond

Abstract This chapter provides a review of the recent developments in water governance around the implementation of market-oriented allocation mechanisms. Following a brief review of China's water scarcity, the chapter sets the theoretical framework for the entire book by reviewing the international debates on water reform. It describes water's characteristics as a resource and the complexity of water use, which have significant implications for shaping water governance. The recent developments in water governance from an international perspective are also examined in this chapter. It covers a number of key concepts which have significant implications for the legal and institutional arrangements of tradable water rights, including interdependency of water use, water as a social and economic good, alternative water property regimes, government versus market in water allocation, and last but not least, climate change adaptation and integrated water resources management.

Keywords China's water scarcity • Water governance debate • Water property regimes • Water allocation mechanisms • Integrated water resources management

2.1 Water Scarcity in China

Leading the world's economic growth, China's development has seriously overstretched its natural resources. It is unsurprising to see that China is one of those countries facing the most severe water crisis around the world. Reflecting the international trend in decreasing water availability and increasing water demand, China's 1.3 billion people and annual economic growth rate of close to 10% (down to 6.5% in 2016–2017) simply need water. Six billion cubic metres of the annual water shortage for urban and industrial demands result in more than RMB 230 billion yuan[1] economic losses of industrial output a year, accounting for 1.62% of the

[1] The Chinese currency is *renminbi* (RMB). *Yuan* is the basic unit of RMB. CNY is the currency code for Chinese Yuan Renminbi. 1 CNY = 0.15 USD (as XE Currency Converter on June 1, 2017).

M. Jiang, *Towards Tradable Water Rights*, Global Issues in Water Policy 18,
DOI 10.1007/978-3-319-67087-4_2

national gross domestic product (GDP).[2] Yet, China faces extremely difficult challenges to satisfy its growing needs for water without sacrificing the sustainability of the resource and the environment.

Despite its world leading rank in total water availability, China is dangerously short of water. With 2813 billion cubic metres of renewable freshwater resources per year, China has the world's fifth largest endowment of fresh water.[3] When it comes to per capita water availability, however, China's water resources become much more modest. Indeed, its 2093 cubic metres per person per year are just one third of the world average (6122 cubic metres).[4] It is estimated that China's population will reach 1.6 billion by 2030, and correspondingly, the per capita water resources will drop further to 1750 cubic metres.[5]

Among China's 660 cities, more than 400 are in short supply of water including 100 of them faced with severe shortage.[6] The situation then worsens as a result of the country's misdistribution of water resources, population density, and economic productivity. China is subject to strong monsoon climate and its precipitation shows great variations from season to season, from year to year, and from region to region. Per capita water availability in different regions of China varies widely: in general, the northern population enjoys much less water availability than their southern counterparts. The wet south has 80.9% of the total available water resources yet only accounts for 34.7% of the country's cultivated land.[7] By comparison, there is only 14.5% of the total water availability for the dry north, which is the home to almost 45% of the nation's population, 59.6% of cultivated land, contributing to 42.8% of the national GDP.[8]

China's rapidly growing energy demand also contributes significantly to its water scarcity. As a large user of water, thermal power plants, 80% of which are coal-fired in China, withdraw water for power plant cooling, accounting for more than 10 % of the national water cap.[9] Moving aggressively towards increased coal production,

---

[2] The 11th Five Year Plan of Water Resources Development of China (China National Development and Reform Commission, Ministry of Water Resources, and Ministry of Construction), 2007, http://www.sdpc.gov.cn/zcfb/zcfbghwb/200706/W020140220601664956788.pdf (accessed October 24, 2016).

[3] World Bank, *World Development Indicators: Table 3.5: Freshwater*, 2014, http://wdi.worldbank.org/table/3.5 (accessed October 24, 2016).

[4] Ibid.

[5] Liping Zhang et al., "Situation and Problem Analysis of Water Resources Security in China," *Resources and Environment in the Yangtze Basin* 18, no. 2 (2009): 116–120.

[6] China Ministry of Water Resources (MWR), *Ministry of Water Resources 2003 Annual Report*, 2004, Beijing: Ministry of Water Resources.

[7] Organisation for Economic Co-operation and Development (OECD) and China Ministry of Agriculture (MOA), *Environment, Water Resources and Agricultural Policies: Lessons from China and OECD Countries*, Paris: OECD, 2006, 77.

[8] Ibid.

[9] World Water Assessment Programme (WWAP), *Water and Energy*, the United Nations World Water Development Report 5, 2014, http://www.unesco.org/new/en/natural-sciences/environment/water/wwap/wwdr/2014-water-and-energy/ (accessed October 24, 2016), 3.

Table 2.1 Nine provincial regions of China with absolute water scarcity

Region	Annual per capita water availability (cubic meters)
Shanghai	143.4
Ningxia	168.0
Beijing	193.2
Tianjin	238.0
Henan	282.6
Shandong	283.9
Shanxi	295.0
Hebei	324.2
Jiangsu	472.0

Data from China National Bureau of Statistics, China 2013 Statistical Yearbook

China's 12th Five Year Plan (FYP) (2011–2015) calls for the creation of 14 large coal industry bases across western China, including water intensive coal mines and coal-fired power plants, intensifying competition over scarce water resources.[10]

The Yellow (Huang) River (*Huanghe*) Basin and the North China Plain are amongst the areas of high water stress and scarcity.[11] In the Yellow River Basin, water scarcity is experienced with 749 cubic metres' per capita water availability.[12] The other two river basins in the north, Hai (*Haihe*) and Huai (*Huaihe*), face even worse water scarcity as their annual water availability is as low as 358 cubic metres and 505 cubic metres per person respectively.[13]

Among 31 provincial level regions (including provinces, autonomous regions, and municipalities directly under the State Council) in mainland China, more than half of them face different levels of water stress and scarcity including nine facing "absolute scarcity" with per capita water resources well below 500 cubic metres[14] (See Table 2.1). Most of these provincial regions with absolute water scarcity fall into the "Three H" Rivers basin area of the North China Plain (Huang-Huai-Hai), where one third of the national GDP is produced with only 7.6% of the national

[10] Ibid, 42.

[11] According to the international definition of water stress and water scarcity, water stress is experienced when annual water availability drops below 1700 cubic metres per person, and it becomes water scarcity when water availability drops below 1000 cubic metres per person per year with 500 cubic metres defined as "absolute scarcity". See WWAP, *Managing Water under Uncertainty and Risk*, the United Nations World Water Development Report 4, 2012, http://www.unesco.org/new/en/natural-sciences/environment/water/wwap/wwdr/wwdr4-2012/ (accessed October 24, 2016), 122.

[12] Hong Yang et al., "The Impact of China's Economic Growth on Its Water Resources: A Regional and Sectoral Assessment," In *Rebalancing and Sustaining Growth in China,* ed. Huw McKay and Ligang Song (Canberra: ANU E Press, 2012), 312.

[13] Ibid.

[14] China National Bureau of Statistics (NBS), *China 2013 Statistical Yearbook*, 2014, http://www.stats.gov.cn/tjsj/ndsj/2013/indexch.htm (accessed October 24, 2016).

water availability.[15] The water scarcity regions in the North China Plain heavily rely on groundwater, accounting for more than 60% of the total fresh water supply.[16] Overexploitation of groundwater has resulted in severe aquifer depletion, making the North China Plain one of the most striking examples of the world for unsustainable groundwater abstraction.

Low water use efficiency exacerbates the already significant gap between water availability and demand in China. As the biggest water using sector, agriculture accounts for 67% of the total water use[17] with its coefficient of effective utilisation as low as 0.50,[18] which indicates that half the water withdrawn for irrigated agriculture is lost before it reaches crops because of evaporation and leakage. The overall economic productivity of water is US$8.0 per cubic metre, compared to a high-income nation's average of US$39.0 per cubic metre.[19]

China's water scarcity challenges also consist of water quality issues. Overdrafts of both surface water and groundwater, increasing wastewater disposal, and low rates of waste water treatment collectively result in water pollution and environmental degradation, which in turn exacerbates water shortage. The Hai Basin is one of the most polluted rivers in China. In 2013, among the river sections in the Hai Basin monitored by the Ministry of Environmental Protection (MEP), only 39.1% was found to range from Grades I to III, which means the water can be used as raw water for potable water supply and direct human contact.[20] Another 21.8% met Grades IV to V, with reduced water quality restricted to industrial use, recreational use with no direct human contact, and agricultural use, while the remaining 39.1% failed to meet Grade V, not suitable for any type of use.[21] Besides rivers running black, lakes and reservoirs, many of which are located in the southern river basins with relatively good water quality, suffer from eutrophication. In 2013, approximately 40% of nationally monitored lakes and reservoirs were under different levels of eutrophication.[22]

While water pollution reduces the already limited water availability and worsens the water shortage in China, human health is imminently threatened by unsafe or

---

[15] OECD and MOA, *Environment, Water Resources and Agricultural Policies*, 77.

[16] Wei Feng et al., "Evaluation of Groundwater Depletion in North China Using the Gravity Recovery and Climate Experiment (GRACE) Data and Ground-based Measurements," *Water Resources Research* 49 (2013): 2110.

[17] China Ministry of Water Resources (MWR) and National Bureau of Statistics (NBS), *Bulletin of First National Census for Water*, Beijing: China Water and Power Press, 2012.

[18] The Water Resources Development Plan (2011–2015) (China National Development and Reform Commission, Ministry of Water Resources, and Ministry of Housing and Urban-Rual Development), 2012, http://www.sdpc.gov.cn/fzgggz/ncjj/nczc/201206/t20120621_487019.html (accessed June 24, 2014).

[19] World Bank. *World Development Indicators*.

[20] China Ministry of Environmental Protection (MEP), *2013 Report on the State of the Environment in China*, 2014, http://www.zhb.gov.cn/zhxx/hjyw/201406/W020140605385940287254.pdf (accessed June 24, 2014).

[21] Ibid.

[22] Ibid.

even poisoned drinking water. By the end of 2010, there were still over 102 million rural people without access to safe drinking water.[23] Although water-related diseases are less common in China compared to other developing countries, some areas where rivers and aquifers suffer from intense pollution have recorded much higher than normal rates of cancer and birth defects, forming notorious cancer villages.[24]

Water quantity and quality problems further lead to severe deterioration of water-based ecosystems. Many northern rivers in China are drying up. In 1972, the Yellow River failed to reach the sea for the first time in the history. From then on, the Yellow River's seasonal desiccation was constant between 1972 and 1998, making 21 years of dry-up out of 27 years.[25] Wetlands around the country are also shrinking dramatically. According to findings of the second national wetland resources survey released in early 2014, China's wetlands have declined by approximately 9%, or by 3.4 million hectares, compared to the findings of the first national survey in 2003.[26]

In the water stressed north where huge gaps exist between water endowment and demand from fast growing cities and water-intensive agriculture, groundwater is heavily relied upon to a highly unsustainable extent. Between 1974 and 2000, groundwater levels of the North China Plain have dropped by approximately 1 m annually.[27] Over-extraction of non-replenishable groundwater causes geological and ecological problems such as land subsidence and seawater intrusion. Among 400 Chinese cities relying on groundwater as a type of water supply, land subsidence is observed in more than 70 cities including some mega cities like Shanghai and Tianjin, where the maximum land subsidence rate exceeds over 2 m.[28]

Due to awareness of the worsening water crisis represented by shortage, pollution, and degrading ecosystems, major efforts have been made to use a range of methods to rescue the water systems. Thanks to generous water transfer, the Yellow River has had its flow back so that since 2000 it has been able to reach the sea. The same method was used to cure two drying inland rivers, Tarim and Hei, whose end lakes have been restored after several decades of dry-up. Nature reserves are set up to protect fragile ecosystems of headwaters of the major rivers, such as Sanjiangyuan National Nature Reserve, which receives massive government funding to restore

---

[23] The 12th Five Year Plan for Rural Drinking Water Safety Projects (China National Development and Reform Commission, Ministry of Water Resources, Ministry of Health, and Ministry of Environmental Protection), 2014, http://www.sdpc.gov.cn/zcfb/zcfbghwb/201402/P020140221360445500781.pdf (accessed July 08, 2014).

[24] The 12th Five Year Plan for Environmental Risk Control of Chemicals (China Ministry of Environmental Protection), 2013, http://www.mep.gov.cn/gkml/hbb/bwj/201302/W020130220539067366659.pdf (accessed July 08, 2014).

[25] Jiwei Chen and Xingmin Mu, "Tendency, Causes and Control Measures on Yellow River Dry-up," *Journal of Natural Resources* 15, no.1 (2000): 31.

[26] Yongli Zhang, *The Deputy Director of the State Forestry Administration's Speech at the Press Conference for the Findings of the Second National Wetland Resources Survey*, http://www.china.com.cn/zhibo/2014-01/13/content_31140091.htm?show+t (accessed July 09, 2014).

[27] Jane Qiu, "China Faces up to Groundwater Crisis," *Nature* 466 (2010): 308.

[28] China Institute of Geo-Environment Monitoring (CIGEM), *China Groundwater Resources*, 2014. http://219.142.70.139/dxs/HydRes.htm#huanjing (accessed July 09, 2014).

and conserve the environment and ecosystems in the headwaters of the Yangtze, the Yellow, and the Lancang (the Mekong River in China's boundary).

Despite all this positive progress, China's water problems remain highly challenging to cope with. One of major exacerbating factors is climate change impacts. Climate change is expected to exacerbate the water crisis as it may have negative effects on the vulnerable freshwater resources.[29] Climate change impacts on water include changes in precipitation patterns, intensity and extremes, snow cover reduction and ice melting, and changes in soil moisture and runoff, resulting in exacerbation of water shortage and pollution.[30] As a result of climate change impacts on other water-related sectors (e.g. agriculture, irrigation and drainage, hydropower, and flood control), water demand and use patterns may be affected, which implies additional stress on water resources.[31] Climate change projections show decreasing trends in annual rainfall in north China and north-east China,[32] expecting intensification of the regions' aridity and water scarcity.

To summarise, this section has provided an overview of China's water scarcity. China's water resource challenges involve water shortage and pollution issues as well as the issue of ecosystem deterioration, each of which presents distinctive challenges for legal and policy development. With a growing population, booming economy, raising standard of living, and changing values of the environment, the competition over scare water resources among different uses becomes more intensified. Adding to these stresses, climate change puts adverse impacts on water resources and other water-related sectors, leading to exacerbation of water problems. Water governance is therefore challenged to be able to balance multiple values of different water uses. It is within this context that an ambitious water reform, "Building a water saving society", has been initiated and carried out in China. This water reform includes attempts to integrate the traditionally fragmented water management systems and implement market-based mechanisms, which will be reviewed in detail in the following chapters. Before that, the rest of this chapter first describes water's characteristics as a resource and the complexity of water use, which both have significant implications for shaping water governance, and then reviews the recent developments in water governance from an international perspective. The review of the international debates on water policy reform in this chapter sets the theoretical framework for the entire book.

---

[29] IPCC, *Climate Change and Water*, IPCC Technical Report VI. Geneva: IPCC, 2008, 7.

[30] Ibid, 13.

[31] Ibid, 33.

[32] Rex Victor Cruz et al., "Asia," In *Climate Change 2007: Impacts, Adaptation and Vulnerability. Contribution of Working Group II to the Fourth Assessment Report of the Intergovernmental Panel on Climate Change*, ed. Martin Parry et al. (Cambridge, UK: Cambridge University Press, 2007), 469–506.

2.2 Challenges Around Water and Water Governance

2.2.1 Understanding Basics of Water and Water Governance

Water needs to be understood within the context of the hydrologic cycle. Through the dynamic, unending and continuous process of the hydrologic cycle,[33] water passes from the atmosphere to the earth and returns to the atmosphere in its three states of liquid, solid and vapour. The atmosphere provides water to the earth through precipitation. Some of it is returned back to the atmosphere by evaporation, transpiration, and sublimation. Other parts left on the earth either drain from land to oceans through surface waters such as rivers, lakes and wetlands, or infiltrate into the ground and replenish aquifers some of which directly reach the oceans.[34] As an essential resource for life, fresh water only accounts for a minor fraction (2.5%) of the total volume of water of the world.[35] Approximately 69% of fresh water is locked up in ice sheets and glaciers, considered inaccessible and not available for use.[36] Another 30% is groundwater.[37] Therefore, holding only 0.3% of the earth's total freshwater resources, surface waters including lakes, rivers, streams, wetlands and reservoirs, represent about 80% of the annually renewable water.[38]

The concepts of "blue water" and "green water" are helpful to further understand the water cycle. Blue waters refer to flows in surface water bodies and aquifers that are directly associated with aquatic ecosystems.[39] In contrast, green water evaporates from plants and water surfaces into the atmosphere as vapour, which supplies terrestrial ecosystems and rain-fed crops from the soil moisture zone.[40] Blue water and green water link multiple users, including humans, ecosystems, and the environment, with each other in the integral hydrologic cycle. Humans use fresh water for various consumptive purposes such as domestic, agricultural, and industrial uses. Apart from these water values that require water diversion, there are other in-stream values such as waste disposal, recreation, and environmental conservation. Ecosystems and the environment need water to sustain other life species and ecosystem health. Within the water cycle, all water uses and users are interdependent and interrelated. Despite the fact that water resources are renewable, water

---

[33] Desheng Hu, *Water Rights: An International and Comparative Study*, London: IWA Publishing, 2006, 10.

[34] WWAP, *Water for People, Water for Life*, the United Nations World Water Development Report 2003, Paris: UNESCO and Berghahn Books, http://www.unesco.org/new/en/natural-sciences/environment/water/wwap/wwdr/ (accessed November 17, 2014), 76.

[35] WWAP, *Water, a Shared Responsibility*, the United Nations World Water Development Report 2, 2006, Paris: UNESCO, http://www.unesco.org/new/en/natural-sciences/environment/water/wwap/wwdr/ (accessed November 17, 2014), 121.

[36] Ibid.

[37] Ibid.

[38] Ibid, 125.

[39] Ibid, 122.

[40] Ibid, 122–123.

availability is limited and subject to variability and uncertainty in space and time. Within a country or a small region, more use in one water using sector often means less water available for another. It is especially true for consumptive uses. In other cases, one type of water use can affect water quality although it does allow reuse by other users. The mobile and fugitive nature of water therefore complicates the coexistence of different users of multiple values.

The nature of water and the multiple values of water use raise challenges for water governance. As many societies around the world are faced with varying water problems such as shortage and pollution, many argue that "the world is in a water crisis",[41] a crisis of water governance or policy.[42] The crisis of governance in sustainable development as well as the global water challenges is now well recognised.[43] Water governance is required to deal with water's variability and uncertainty, to balance multiple values of different uses, and to avoid and/or solve conflicts that are likely to intensify among increasingly competing water users.

According to the definition by the United Nations Development Programme (UNDP), governance refers to "the exercise of political, economic and administrative authority in the management of a country's affairs at all level".[44] It "comprises the complex mechanisms, processes and institutions through which citizens and groups articulate their interests, mediate their differences and exercise their legal rights and obligations".[45] Governance is more than just government, but includes the private sector and civil society, reflecting interaction and negotiation among the three important domains of institutions and organisations.[46] The concept of water governance has been used differently in various contexts. In some cases, water governance can be used to refer to the actual water policy in place.[47] However, water governance involves much more than just water policy. In addition to policy itself, governance is also about the exercise of power in policy making and whether or not to implement particular policies.[48] In essence, water governance refers to "the range of political, social, economic and administrative systems that are in place, which directly or indirectly affect the use, development and management of water resources

[41] WWAP, *Water for People, Water for Life*, 5.

[42] WWAP, *Water, a Shared Responsibility*, 45.

[43] See for example, United Nations Development Programme (UNDP), *Discussion Paper: Governance for Sustainable Development: Integrating Governance in the Post-2015 Development Framework*. New York: UNDP, 2014. Also, Margot Hill, *Climate Change and Water Governance: Adaptive Capacity in Chile and Switzerland* (Dordrecht, Heidelberg, New York, London: Springer, 2012).

[44] UNDP, *Governance for Sustainable Human Development*, A UNDP policy document. New York: UNDP Management Development and Governance Division and Bureau for Policy and Programme Support, 1997, 2.

[45] Ibid.

[46] UNDP, *Reconceptualising Governance*, Discussion paper 2. New York: UNDP Management Development and Governance Division and Bureau for Policy and Programme Support, 1997, 13–18.

[47] WWAP, *Water, a Shared Responsibility*, 48.

[48] Ibid.

and the delivery of water services at different levels of society".[49] Water governance therefore encompasses mechanisms, processes, institutions, relationships and networks through which water resources are managed, negotiated, and allocated. By allocating water resources among multiple water using sectors, water governance is about making choices, decisions and trade-offs.[50]

Population growth, urbanisation, economic development, and diminishing water supply due to the combination of both natural climate variability and human activities (e.g. water pollution) are already stressing water availability. The recent manifestation of climate change creates new challenges on sustainable development of water resources and water governance. Climate change impacts on hydrological resources and patterns include alterations in the timing, spatial distribution, quality and quantity of water resources. Moreover, climate change has complex interactions with other drivers of change through for example food and energy production and consumption, which presents as an exacerbating factor affecting water governance.

Indeed, climate change brings water governance into a new era of nonstationarity and uncertainty.[51] Stationarity assumes that "natural systems fluctuate within an unchanging envelope of variability",[52] based on which water governance was historically designed. In the context of increasingly uncertain future conditions of climate and water resources, water governance can no longer rely on the assumption that the natural systems will "replicate the relatively stable conditions of the past",[53] and may have to make adjustments and/or shift focus to be more able to deal with nonstationarity and uncertainty. How water governance responds to increasing water scarcity, variability and uncertainty is therefore a key global public policy challenge. Water reforms have been observed globally that tell interesting stories of mixed success and failure with a common desire to achieve balance in multiple values of water use. Laws, regulations, policies and actions, institutions, and networks among organisations are being revisited and examined to underpin good water governance.

2.2.2 *Water as a Human Right vs. Water as an Economic Good*

Water is a social and economic good. It is an indispensable social resource upon which humans rely in order to survive. As water becomes scarcer, its economic value among its competing uses is gaining wider recognition in international forums. Water's social and economic values are reflected in water governance debates on treating water as a human right versus allocating and pricing water as an economic

---

[49] Ibid, p. 47.

[50] Ibid.

[51] P.C.D. Milly et al., "Stationarity Is Dead: Whither Water Management?" *Science* 319 (2008): 573–574.

[52] Ibid, 573.

[53] Hill, *Climate Change and Water Governance*, 7.

good. Safe drinking water and sanitation are basic human needs. Unfortunately, despite substantial efforts in meeting the Millennium Development Goals (MDGs) for drinking water and sanitation since 1990, there are still more than 700 million people without access to safe drinking water, 2.5 billion people without an improved sanitation facility and of these one billion still practising open defecation.[54]

While the basic needs for water in some cases are failed due to physical water scarcity, in many other cases, it is more an institutional and management failure in water supply. The concept of water as a human right emerged to address this issue. The literature has well argued the importance of explicitly acknowledging an independent human right to water, as opposed to subordinate to other primary human rights such as health or life, especially at the domestic law level.[55] However, the explicit recognition of an independent human right to water in international law has been a long struggle.[56] In 2002, the United Nations (UN) Committee on Economic, Social and Cultural Rights (CESCR), which monitors the covenant of the same name (ICESCR), adopted the General Comment No. 15 on the right to water. It states that "the human right to water entitles everyone to sufficient, affordable, physically accessible, safe and acceptable water for personal and domestic uses".[57] Adding both legal and policy dimensions to the debate on the human right to water, General Comment No. 15 states that the right to water, like any human rights, imposes obligations on state parties. General Comment No.15 is non-binding but authoritative interpretations of ICESCR rights and obligations, being a reference by various international bodies when deciding whether a state has met its obligations under the covenant.[58]

In 2010, the UN general assembly passed a resolution (A/RES/64/292) explicitly recognising the human right to water and sanitation. It acknowledges "the right to safe and clean drinking water and sanitation as a human right that is essential for the full enjoyment of life and all human rights".[59] Two months later in the same year, the UN Human Rights Council adopted Resolution A/HRC/RES/15/9 affirming that the rights to water and sanitation are part of existing international law, legally

[54] World Health Organisation (WHO) and United Nations Children's Fund (UNICEF), *Progress on Drinking Water and Sanitation (2014 Update)*, Geneva, Switzerland: WHO and UNICEF, 2014, Forward.

[55] Anonymous, "Water Prices for the Priceless?: Implementing the Justifiability of the Right to Water," *Harvard Law Review* 120 (2007): 1076–1078. Also see Erik B. Bluemel "The Implications of Formulating a Human Right to Water," *Ecology Law Quarterly* 31 (2004): 957–1006.

[56] Farhana Sultana and Alex Loftus. eds. *The Right to Water: Politics, Governance and Social Struggles* (USA and Canada: Earthscan, 2012).

[57] United Nations Economic and Social Council, *General Comment No. 15 on the Right to Water*, 2002, http://www2.ohchr.org/english/issues/water/docs/CESCR_GC_15.pdf (accessed May 2, 2017), Paragraph 2.

[58] Bluemel, "The Implications of Formulating a Human Right to Water".

[59] United Nations General Assembly, *Resolution Adopted by the General Assembly on 28 July 2010*, A/RES/64/292, United Nations, 2010, http://contrattoacqua.it/public/upload/1/2/tab_elms_docs/1404320102n0947935.pdf (accessed May 2, 2017), 2.

binding upon states.[60] It calls on states to develop appropriate tools and mechanisms to progressively achieve the full realisation of human rights obligations related to access to safe drinking water and sanitation.[61] As a result of the "extraordinary breakthrough in the international struggle"[62] for the right to water represented by the two resolutions, every UN member state is now required to prepare a Plan of Action for the Realisation of the Right to Water and Sanitation to meet the obligations to respect, protect, and fulfil the right to water, and to report to the UNCESCR on its performance in this area.[63]

Mirroring the political difficulties in the explicit recognition of an independent human right to water in international law, the domestic implementation of a right to water has been through political, legal and social struggles around the world. Except for a few countries where a legally enforceable right to water exists based on domestic law (e.g. South Africa), most countries have not yet made formal legal commitments to acknowledging a human right to water.[64] Rather, the right to water has been mostly protected by expansive interpretations of constitutional fundamental rights, especially the right to life.

On the other side of the debate on human rights to water is the view of water as an economic good, mainly framed by the 1992 Dublin Statement on Water and Sustainable Development (the fourth principle) and Chapter 18 of UN's Agenda 21. Managing water as an economic good has significant implications for water governance, especially in water allocation and water pricing. Traditionally, water has been supplied as a public and sometimes even free resource. Water users have often been charged only a proportion of the costs of water supply thanks to public finance subsidies, if it is not at all free of charge. The principle of full cost recovery associated with water as an economic good therefore raises concerns that commercialisation, privatisation, and commodification of water may lead to water being unaffordable or inaccessible for the poor. To this end, water governance again meets with the challenges of how the cost of water (costs of water supply, opportunity costs, and externalities) is compensated, how the value of water (economic, environmental, social, and cultural) is appreciated, and how the price of water reflects its cost and value.

The 1992 Dublin Statement interestingly expresses the double challenge of water as both a social and economic good. Its fourth principle states that "water has an economic value in all its competing uses and should be recognised as an economic good", it then further elaborates that "within this principle, it is vital to recognise first the basic right of all human beings to have access to clean water and sanitation at an affordable price". It may sound straightforward, but debates remain in what

[60] Sultana and Loftus, *The Right to Water,* xv.

[61] United Nations Human Rights Council, *Resolution Adopted by the Human Rights Council, 15/9 Human Rights and Access to Safe Drinking Water and Sanitation*, A/HRC/RES/15/9, United Nations, 2010.

[62] Sultana and Loftus, *The Right to Water,* xv.

[63] Ibid.

[64] WWAP, *Water, a Shared Responsibility,* 65.

takes for water governance to do that. While water shortages are increasingly experienced in many countries, the concepts of a human right to water and water as an economic good seem to come into conflict with each other. These two approaches tend to take different, sometimes opposite directions towards the roles of government and market in water governance. Water has been traditionally provided, management and allocated with substantial involvement of government as a public good. The economic dimension of water, on the other hand, favours market force in water allocation. As a key element of water governance, water allocation mechanisms present a spectrum of institutional and legal alternatives in which government and market play different roles.

2.2.3 *Alternative Property Regimes as a Water Policy Instrument*

According to Bromley's property regimes framework, property rights are used as an instrument to govern water use and allocation.[65] Property is "a social relation that defines the property holder with respect to something of value against all others".[66] Property is therefore a "social instrument", and "particular property regimes are chosen for particular purposes".[67] Alternative property-rights regimes include: non-property (open access), common property, state property, and private property, representing a spectrum with private property at one extreme and public property at the other, and various mixes of the two in between.[68] Although lawyers and economists adopt different definitions to property rights, they share a common idea that a property right mush have "social recognition, enforcement and protection of the holder's powers given under the right".[69] A property right refers to "the ability to do something which is recognised directly or indirectly in law, to the extent that the law can be expected positively to protect that ability or negatively to prevent others from interfering with that ability".[70] A water property right specifically involves the following elements[71]:

[65] Daniel W. Bromley, *Environment and Economy: Property Rights and Public Policy* (Oxford, UK; Cambridge, USA: Blackwell, 1991).

[66] Ibid, 2.

[67] Ibid, 3.

[68] Ibid, 23. Also see ACIL Tasman and Freehills, *An Effective System of Defining Water Property Titles*, Research Report to the Australian Government Department of Agriculture, Fisheries and Forestry, and Land & Water Australia, Canberra: Land & Water Australia, 2004, 17.

[69] Anthony Scott and Georgina Coustalin, "The Evolution of Water Rights," *Natural Resources Journal* 35, no. 4 (1995), 822.

[70] Ibid, 824.

[71] ACIL Tasman and Freehills, *An Effective System of Defining Water Property Titles,* 15–16. Also see Edella Schlager and Elinor Ostrom, "Property Rights Regimes and Natural Resources: A Conceptual Analysis," *Land Economics* 68, no. 3 (1992): 250–251.

- Access: the right to enter a defined physical water system and enjoy non-subtractive benefits, such as navigation and swimming;
- Withdrawal: the right to obtain resource units or products of a water system, such as abstract water for off-stream uses or to catch fish;
- Management: the right to regulate internal use patterns and transform the resource by making improvements, such as dam and reservoir construction and water transfer diversion;
- Exclusion: the right to determine who will have the rights of access and withdrawal, and how those rights may be transferred;
- Alienation: the right to sell or lease management and/or exclusion rights.

Accordingly, property rights can be seen as a bundle of individual rights among which the right of access and the right of withdrawal are operational-level rights, and the rights of management, exclusion and alienation are more extensive rights authorising participation in collective-choice actions.[72] Such specification of water property rights makes a distinction between property rights and ownership. Ownership is a property right in an absolute sense, but one can hold a water property right without being the owner of the resource. The right of alienation defines ownership, except which property right holders other than owners can have either one or all of the rights of access, withdrawal, management, and exclusion.[73]

In the non-property (open access) regime, there is no defined group of users or owners over water resources. Water resources are available to anyone, for which there are no property rights but only possession in this regime.[74] Water "belongs to no-one until it is in someone's physical possession",[75] and will belong to the one who first exercises capture. Although the open access water is very likely to be overused, the open access regime may be appropriate to ensure equitable access to water for all.[76] In other words, the open access regime can be a conscious policy choice rather than ignorance of the "institutional vacuum" (the absence of rules defining property rights).[77] The common property regime defines a group of members who have the property right over the resource and excludes non-members from the use.[78] Common property right can be seen as private property for the group of co-owners, within which individual members are able to use segments of the resource held under it.[79] Compared to the common property right holders who only hold use (usufruct) rights with no rights of alienation, the private property regime

[72] Ibid.

[73] Ibid.

[74] Bromley, *Environment and Economy*, 32.

[75] Ibid.

[76] Elinor Ostrom, "Private and Common Property," *Encyclopedia of Law and Economics* II (2000), 336.

[77] Ibid. Also see Bromley, *Environment and Economy*, 30, 148.

[78] Bromley, *Environment and Economy*, 31.

[79] Ibid, 25.

gives individuals the full range of rights over the resource.[80] Under the state property regime, the state, rather than individuals and groups, holds the ownership of water resources and controls water use. Individuals and groups may be able to make use of water based on usufruct rights. The state may directly manage and control the use through governmental agencies.[81]

It is interesting to observe that water property rights regimes can be vastly different from country to country or even from jurisdiction to jurisdiction within a single country. The choice of a certain regime depends on the local context and works for specific purposes. There is no one single perfect solution to suit all circumstances in water use. Water policy observers always need to be aware of and appreciate the diversity of water rights regimes across jurisdictions, even those sharing a same legal system. For instance, great variation in water rights can be found in the United States (US) and Australia despite their historical origin of English common law.[82] The eastern and westerns states of the US have developed significantly distinct water rights law as a result of different history in response to the relative scarcity of local water resources.[83] Water availability and water use play a critical role in shaping water property rights regimes. The choice of institutional arrangements depends on different ecological circumstances and cultural, political and economic contexts.[84] Their complexity is also reflected in the mixture of alternative water property rights regimes. These contributing factors influence the evolution of water property rights regimes as they change. As water becomes scarcer, increasing competition among water users would call for more defined water property rights. Open access regimes may be transformed to some form of property regimes. Similarly, one type of property regime (private, common, or state) could be transformed to another type. The institutional arrangements during such evolution are more likely to represent a combination of multiple property regime components rather than a choice of one single regime.

2.2.4 The Water Allocation Mechanisms Debate

The evolution of water rights regimes is often related to the debate on the institutional arrangements of water allocation mechanisms. In the spectrum of property rights regimes, it involves different levels of government intervention in water management and allocation.[85] The state property regime requires substantial government

---

80 Ibid.

81 Ibid.

82 Richard A. Epstein, "The Historical Variation in Water Rights," In *The Evolution of Markets for Water: Theory and Practice in Australia*, ed. Jeff Bennett (Cheltenham, UK, Northampton, MA, USA: Edward Elgar, 2005), 24–37.

83 Terry L. Anderson et al., *Tapping Water Markets* (New York: Routledge, 2012), 26.

84 Bromley, *Environment and Economy*, 157.

85 ACIL Tasman and Freehills, *An Effective System of Defining Water Property Titles*, 19.

control or agency administration. At the other end of the private property regime lies the promotion of market and its efficacy for more efficient and sustainable water use in the context of the increasing scarcity of water resources. Self-governance of water users (user group management) in between based on common property presents the third institutional option.[86] Mirroring the property rights regimes, the institutional alternatives for water allocation can also be viewed as a spectrum from complete government control at one extreme through a mixture of market and government allocation in the middle, to predominant market allocation at the other extreme.[87] In addition, user-based allocation also plays a part, which may be combined with market and/or government allocation. Such combination of institutions may be necessary and more effective to address the complexity of water use. Water reform is therefore a process of exploring a balanced mix of water allocation mechanisms suitable for a specific local context as the level of water scarcity varies.

Government allocation as an institutional alternative is different from government involvement. As opposed to direct control over water and water use under the "government allocation institution", government is involved in water allocation under the other two institutions (user-based allocation and market allocation).[88] Even the purest market allocation requires government support and intervention, which establishes regulatory frameworks for users' self-governance and market exchange.[89] Government allocation here refers to a form of water institution in which water is owned by the state, with "mandate authorising bureaucratic agencies to control water directly".[90] With such direct control, water allocation is primarily based on the professional norms (such as irrigation schedules and reservoir operation rules) and bureaucratic procedures with little user participation and communication.[91] Government agencies are responsible not only for building and operating water supply systems, but also for carrying out "a range of regulatory roles, allocating rights, resolving disputes, and reallocating water".[92] As a result, there is "little or no formal specification of water rights of the ultimate users" in such agency-controlled systems.[93]

From the perspective of water as a public good, water is viewed as indivisible and non-excludable. "Indivisibility" means that water cannot be broken up and its benefits sold individually with joint or no rival in supply, whilst "non-excludability"

[86] Bryan R. Bruns and Ruth Meinzen-Dick, "Frameworks for Water Rights: An Overview of Institutional Options," In *Water Rights Reform: Lessons for Institutional Design,* edited by R.B. Bruns, C. Ringler, and R. Meinzen-Dick (Washington, D.C.: International Food Policy Research Institute, 2005), 7.

[87] Ariel Dinar et al., *Water Allocation Mechanisms: Principles and Examples*, World Bank Policy Research Working Paper, No. 1779, 1997, 3.

[88] Bruns and Meinzen-Dick, "Frameworks for Water Rights," 9.

[89] Dinar et al., *Water Allocation Mechanisms*, 3.

[90] Bruns and Meinzen-Dick, "Frameworks for Water Rights," 9.

[91] Ibid.

[92] Ibid.

[93] Ibid.

means that water's benefits cannot be excluded from those who do not provide for it.[94] One's appreciation of a river view and its wildlife does not subtract from the possibility of others having the same enjoyment. Similarly, water pollution control and maintenance of water ecosystems benefit all users. It is therefore hard to govern these ecosystem services through market as there are no discrete groups of providers and beneficiaries.[95] The logic choice of allocation mechanisms would then come to government. Whereas water quality displays more of the attributes of a public good, water quantity has more common-pool characteristics. Common-pool resources are defined as natural or man-made resources of which use is subtractable, but exclusion is difficult and costly, although it is not completely impossible.[96] Lakes, rivers, and aquifers all have such characteristics, allowing multiple users to benefit from their use, while at the same time, there is rivalry in its consumption. Both the public good and common-tool characteristics of water make it difficult to manage and allocate at the individual level, which justifies the institutional rationality of government allocation.

The notion of water as a human right also tends to favour government allocation for water resources that intends to promote equity in water allocation, and assure human societies with a minimum supply of basic water needs. The protection of the poor in society remains the responsibility of government and is unlikely to be taken care of by market. There are some other public values for which water serves, such as sustaining the environment or preventing floods. These values again need government to take control of water resources. Due to these and other reasons mentioned above, government allocation of water has been employed as the main mechanism in many countries.[97] While the government allocation mechanism has worked well in many cases, incentive structures within government could generate "government (public) failure".[98] Some of the most common "ills of the public sector"[99] include: rent seeking in both economic and socio-political forms,[100] incentives that are not related to performance and efficiency,[101] special-interest effect (politically powerful

[94] ACIL Tasman and Freehills, *An Effective System of Defining Water Property Titles*, 17.

[95] James Salzman, *Creating Markets for Ecosystem Services: Notes from the Field*, Duke Law School Science, Technology, and Innovation Research Paper No. 2., 2005, 113.

[96] Elinor Ostrom, *Governing the Commons, the Evolution of Institutions for Collective Action* (Cambridge, New York: Cambridge University Press, 1990).

[97] Dinar et al., *Water Allocation Mechanisms*, 3. Also see Fredrik Segerfeldt, *Water for Sale: How Business and the Market Can Resolve the World's Water Crisis* (Washington D.C.: Cato Institute, 2005), 80.

[98] Terry L. Anderson, "The Water Crisis and the New Resources Economics," In *Water Rights: Scarce Resource Allocation, Bureaucracy, and the Environment*, ed. Terry L. Anderson (San Francisco, California; Cambridge: Pacific Institute for Public Policy Research; Mass, Ballinger Pub. Co., 1983), 5.

[99] C. J. Perry, et al., *Water as an Economic Good: A Solution, or a Problem?* International Irrigation Management Institute Research Paper, 14, 1997, 7.

[100] Dinar et al., *Water Allocation Mechanisms*, 7.

[101] Ibid. Also see Anderson, "The Water Crisis and the New Resources Economics," 5–6.

interests and their clients dominate the policy arena and receive political favours),[102] and short sightedness effect.[103] In practice, government allocation mechanisms often cause the problem of waste and misallocation of water, sometimes through subsidies and irrational water pricing.

Based on these government failures, some writers come to the conclusion of privatising water allocation for improved economic efficiency. Water markets and tradable water rights are the two main forms of such privatisation.[104] From an economics perspective, using one additional unit of water in one sector should theoretically produce the same marginal benefits when used in any other sector in order to maximise social welfare. If it is not the case, the sector that produces higher benefits should be allocated more water for its more efficient use. The function of market allocation to direct water to economically higher valued uses has been widely recognised. As mentioned earlier, the right of alienation is one part of the bundle of water property rights, which involves decision making relating to resource allocation or transfer. In a government allocation system, water users often only have the rights of access and withdrawal while the right of alienation is held by government agencies. In contrast, a market allocation system is featured with voluntary transactions between users. Aware of the opportunity costs, a water right holder may consider using water themselves or transferring the right to others. If the transfer is more profitable than one's own use based on the fact that the other user values water higher, the water right holder may decide to alienate the right. To meet their own need for water, the water right holder may improve water use efficiency first and then sell or lease the surplus to others.

Water markets and tradable water rights are two different yet overlapping concepts.[105] Water markets can exist in an informal manner in the absence of a formal definition of water rights.[106] In practice, informal water markets most commonly operate at the local level among similar uses of water such as those between tube well owners and other farmers in Pakistan and Bangladesh.[107] Responding to government failures in water allocation, these informal water markets help irrigators meet their needs in water shortage. However, informal water markets tend to be limited in short-term and in-sector transactions due to difficulty in enforcing such contracts and low security of water access without formal legitimisation for water transfer. Tradable water rights, on the other hand, require "both the formal definition

---

[102] Anderson, "The Water Crisis and the New Resources Economics," 5–6.

[103] Ibid, 5.

[104] For recent observations on water markets, see K. William Easter and Qiuqiong Huang, eds. *Water Markets for the 21st Century: What Have We Learned?* (Dordrecht: Springer, 2014).

[105] Many writers have indicated this difference. For example, Easter and Huang, *Water Markets for the 21st Century*. Also Perry et al., *Water as an Economic Good,* 7. Also, Bryan R. Bruns and Ruth Meinzen-Dick, *Negotiating Water Rights* (London: ITDG Publishing, 2000), 41.

[106] Bruns and Meinzen-Dick, *Negotiating Water Rights*, 41.

[107] Ibid. Also see Syed Z. Sadeque, "Nature's Bounty or Scarce Commodity: Competition and Consensus over Groundwater Use in Rural Bangladesh." In *Negotiating Water Rights*, ed. Bryan R. Bruns and Ruth Meinzen-Dick (London: ITDG Publishing, 2000), 269–291.

of entitlements, and the specification of the conditions under which the entitlement may be traded".[108] They are central to the development of formal water markets, more likely to involve intersectoral transfers such as irrigation water transferred to urban or industrial uses. Water trading practices in Australia, Chile, and the western US are prominent examples of formal and highly organised water markets.

Water rights trading practices in those jurisdictions have taken various approaches to government intervention in water allocation to suit their particular local contexts. Chile has been "the world's leading example of the free-market approach to water law and water resources management",[109] in which water rights can be freely bought, sold, mortgaged, and transferred like real estate.[110] As an extreme example, Chile is "the only country in the world whose water legislation includes no obligation to use water rights".[111] Chilean water rights holders are free to make changes in water use types and methods with no need for administrative approval, and they face neither penalty nor risk of cancellation of water rights for lack of use.[112] Other jurisdictions with water trading markets in place usually put restrictions on the exercise and transfer of water rights. In the western US, the exercise of water rights is subject to the rule of beneficial use, although water rights are treated as private property in perpetuity. Compared to Chile and the western US where water rights are perpetual with high-level security similar to other private property rights, Australia takes a different approach to water rights with higher risk of attenuation because government keeps the power to review water rights in association with planning processes. Rather than being freely transferable, water rights transfer in Australia in most cases is subject to the public agency's approval.[113]

As mentioned previously, some level of government involvement is required even for the most predominant market allocation mechanisms. In Chile, for example, the General Water Directorate is responsible for granting requests for new water rights, keeping official registries of water rights, and governing private water users' organisations.[114] In many other cases of water rights trading, more government intervention is observed in water allocation and management including two key forms: defining and allocating the initial water rights, and establishing the "rules of the game", which is the institutional and legal framework for trade.[115] Because of

[108] Perry et al., *Water as an Economic Good*, 7.

[109] Carl J. Bauer, *Siren Song: Chilean Water Law as a Model for International Reform* (Washington & Great Britain: Resources for the Future, 2004), 1.

[110] Carl J. Bauer, *Against the Current: Privatization, Water Markets, and the State in Chile* (Boston & London: Kluwer Academic, 1998), 34.

[111] Ibid, 46.

[112] Ibid.

[113] Mark Bartley et al., *Trading in Water Rights: Towards a National Legal Framework* (Sydney and Brisbane: Phillips Fox; Queensland University of Technology, 2004), 82.

[114] Bauer, *Against the Current*, 35–36.

[115] Paul Holden and Mateen Thobani, *Tradable Water Righs: A Property Right Approach to Resolving Water Shortages and Promoting Investment*, World Bank Policy Research Working Paper 1627, 1996, 6.

water's unique characteristics, its market allocation cannot be the same as that of other economic goods as more government intervention may be necessary to address externality issues and risk of market failures, which are linked with the nature of water cycle and interdependence among users. Not surprisingly, the free water market of Chile has been criticised for its weaknesses in addressing a range of water management issues including social equity, environmental protection, and river basin management.[116] After a few decades of international water reform, government versus market in water allocation remains a heated debate.[117] At the centre lies the need for a balanced combination of market and government water allocation institutions.

2.2.5 Climate Change Adaptation and Integrated Water Resources Management (IWRM)

Water stakeholders have long dealt with stresses and changes relating to climate variability. However, the changes in water availability and variability associated with climate change will require more adaptive responses to exacerbated uncertainty and stress. According to the definition by the Intergovernmental Panel on Climate Change (IPCC), adaptation to climate change refers to a process of adjustment to actual or expected climate, and its effects, in order to moderate harm or exploit beneficial opportunities.[118] Researchers and policy makers historically paid more attention to mitigation of climate change, but progress in reducing greenhouse gas emissions over the past two decades has remained modest. Even with radical emissions reductions, further climate change over the next few decades is predicted inevitable,[119] leading to widespread recognition that it is imperative for all societies and economic sectors to adapt to climate change impacts. Freshwater resources in many countries around the world are heavily pressured by population growth, economic development, rising standards of living, changing values of the environment, and on top of all these stresses, climate change. In this context, adaptation to climate change becomes a critical part of water governance. Water governance needs to be more adaptive, flexible, and integrative, to be able to respond to future climatic stress and uncertainty.[120]

[116] Carl J. Bauer, "Water Conflicts and Entrenched Governance Problems in Chile's Market Mode,l" *Water Alternatives* 8, no.2 (2015): 147–172. Also see Bauer, *Siren Song*, 2.

[117] Anderson et al., *Tapping Water Markets*.

[118] IPCC, *Special Report on Managing the Risks of Extreme Events and Disasters to Advance Climate Change Adaptation (SREX), Summary for Policy Makers*. Geneva: IPCC, 2011, 3.

[119] IPCC, "Summary for Policymakers," In *Climate Change 2013: The Physical Science Basis. Contribution of Working Group I to the Fifth Assessment Report of the Intergovernmental Panel on Climate Change*, ed. by T. F. Stocker et al. (Cambridge and New York: Cambridge University Press, 2013).

[120] Hill, *Climate Change and Water Governance*.

Climate change adaptation involves a combination of technological, behavioural, and managerial changes, in which governance, institutions, law and policy in particular, can play a significant role to underpin and facilitate such changes.[121] Legal and policy changes to water property regimes, allocation mechanisms, and their implementation and administration could be one of the possible adaptive institutional responses to climate change.[122] The development of water markets and tradable water rights is indeed one of the important potential adaptation tools.[123]

As the imperative of adaptation to climate change becomes increasingly recognised, there is a recent emergence of theory, policy, and practice in promoting integrated water resources management (IWRM) as a tool for adaption to climate change.[124] Against the traditional fragmented sectoral water management, IWRM has been widely accepted and promoted as a management paradigm for sustainable water use and management.[125] Since Chapter 18 of Agenda 21 and the Dublin Principles formulated in the 1992 International Conference on Water and the Environment in Dublin introduced the concept of IWRM, there has been universal support to it in the international community. Efforts have also been made in its implementation at a domestic level. Without a commonly agreed definition, IWRM can be used in many different ways. The Global Water Partnership (GWP) has proposed a framework which defines IWRM as "a process which promotes the coordinate development and management of water, land and related resources, in order to maximise the resultant economic and social welfare in an equitable manner without compromising the sustainability of vital ecosystems".[126]

The mobile and fugitive nature of water complicates the coexistence and interdependence of water users, which is the basis of the IWRM concept. Based on the understanding of the integral water cycle and water as a social and economic good, IWRM promotes integration of water management: both natural system integration and human system integration.[127] The GWP's IWRM framework comprises four major elements, namely integration of river basin (catchment) management; integration of water management and other related resources management; integration of social, economic, and environmental values of water; and integration of all stakeholders in water management. While the first two elements focus more on the

[121] Tim Bonyhady et al., eds. *Adaptation to Climate Change: Law and Policy* (Sydney: The Federation Press, 2010).

[122] Sheila M. Olmstead, "Climate Change Adaptation and Water Resource Management: A Review of the Literature," *Energy Economics* 46 (2014): 500–509.

[123] Ibid.

[124] Animesh K. Gain et al., "Can Integrated Water Resources Management Increase Adaptive Capacity to Climate Change Adaptation? A Critical Review," Journal of Water Resource and Protection 5 (2013): 11–20. Also see Hill, *Climate Change and Water Governance*.

[125] WWAP, *Managing Water under Uncertainty and Risk*, the United Nations World Water Development Report 4. Paris: UNESCO, 2012, http://www.unesco.org/new/en/natural-sciences/environment/water/wwap/wwdr/ (accessed November 17, 2014), 206.

[126] Global Water Partnership Technical Advisory Committee, *Integrated Water Resources Management*, TAC Background Papers, No. 4, 1998.

[127] Ibid.

natural system integration and the last two on human system integration, it is highlighted that integration must occur both within and between the natural and human systems, taking into account variability in time and space.[128]

Recent studies on IWRM and climate change adaptation have acknowledged opportunities and challenges of IWRM as a climate change adaptation framework for the water sector.[129] By emphasising on collaborative governance and recognising multiple values of water, IWRM is viewed as a means to enhance a system's resilience to cope with climate change impacts on water resources.[130] Meanwhile challenges exist in mainstreaming long-term future focus of climate change adaptation dealing with high-level uncertainty, into IWRM mainly focused on current and historic issues.[131] How IWRM is implemented in practice and how its implementation positively contributes to adaptive capacity also remain questionable.[132] Nevertheless, while experiencing substantial growth in the body of knowledge in climate change adaptation and water governance, our understanding of the linkages between IWRM and climate change adaptation is still limited. To address this gap, the rest of this section aims to establish explicit linkages between climate change adaptation and the four elements of IWRM. Such connections are expected to set up a context for governance integration in which the development of tradable water rights takes place.

River Basin as Water Management Unit The hydrological, geomorphological, and ecological interactions of water resources relate to the basin as their basic functioning unit area rather than political or administrative boundaries, which makes the basin a logical focus for water management.[133] River basins, especially the major ones, are the centre of human civilisation, where how water is used by one has significant consequences for another. As GWP highlighted in its IWRM framework, integration should occur within and between natural systems and human systems. For the integration of natural systems, it involves the river basin's operational unity of water, land, and ecology.[134] The integration of human systems involves multiple uses and functions of the river, their implications for each other, and how they are planned and managed. The evolution of IWRM has seen the development of river basin authorities, the coordination and collaboration between related government

---

[128] Ibid.

[129] For example, see Fulco Ludwig et al., "Climate Change Adaptation and Integrated Water Resources Management in the Water Sector," *Journal of Hydrology* 518, Part B, (2014): 235–242.

[130] Hill, *Climate Change and Water Governance*, 25.

[131] Ludwig et al., "Climate Change Adaptation and Integrated Water Resources Management in the Water Sector".

[132] Josselin J. Rouillard et al., "Evaluating IWRM Implementation Success: Are Water Policies in Bangladesh Enhancing Adaptive Capacity to Climate Change Impact?" *International Journal of Water Resources Development* 30, no. 3, (2014): 515–527.

[133] Peter W. Downs et al., "How Integrated is River Basin Management?" *Environmental Management* 15, no. 3 (1991): 299–309.

[134] Ibid.

agencies and other policy networkers, and the production of comprehensive river basin plans.[135] These institutional strategies highlight the recognition of river basins as the management unit aimed at unified management of rivers' different functions such as water supply, flood control, transport, waste disposal, conservation, and so on. Water planners and managers have long dealt with climate variability and its implications for water demands, but mainly based on the conventional assumption of stationarity.[136] As the impacts of climate change on water availability and demand intensify, water planning and management are challenged by the increasing uncertainty of future conditions. Adaptation to climate change can be divided into autonomous adaptations and planned adaptations.[137] Autonomous adaptations are not deliberately designed to cope with climate change but are responses to meet other changes in demands, objectives and expectations.[138] Planned adaptations, on the other hand, constitute conscious responses to climate stimuli and deliberate policy decisions taking climate change into account.[139] With the currently remaining limited implementation around the world, planned adaptations will be increasingly needed for the water sector to build its adaptive capacity. Legal and institutional frameworks of integrated river basin management have the potential to contribute to incorporating climate change issues into the roles of river basin authorities and river basin plans. While adaptation requires collective effort of all the stakeholders from organisations to individuals, the coordinative and collaborative framework that river basin management offers is expected to facilitate the adaptation decision-making processes.

Integrating Water Resources Management with Other Related Resources Management During the hydrological cycle, water travels between surface and groundwater systems, between land and oceans, and between the forms of blue and green water. It means that how the terrestrial systems and the coastal zones are used and managed may have direct implications for water quantity and quality. Mechanisms are required to ensure full inclusion of all types of water resources, not only rivers, lakes, aquifers, and wetlands, but also relevant transitional and coastal waters. Agriculture is the largest consumer of water, globally contributing to about 70% of the total consumption of fresh water.[140] An expansion in irrigated land is expected in the future to meet the needs of food supply for the growing population. Agricultural appropriation of water, however, is increasingly challenged by the development of urbanisation and industrialisation. In addition, the return flows with pollutants from irrigation land may affect water quality in the downstream waters.

[135] Nigel Watson, "Integrated River Basin Management: A Case for Collaboration," *International Journal of River Basin Management* 2, no. 4 (2004): 243–257.

[136] Milly et al., "Stationarity is dead," 573–574.

[137] IPCC, *Climate Change and Water, 48.*

[138] Ibid.

[139] Ibid.

[140] WWAP, *Water in a Changing World*, the United Nations World Water Development Report 3, Paris: UNESCO and Earthscan, 2009, http://www.unesco.org/new/en/natural-sciences/environment/water/wwap/wwdr/ (accessed November 17, 2014), 16.

Water resources, both in terms of quantity and quality, are critically influenced by agriculture, change in land use and land cover, and pollutant emissions.[141] Consequently, interaction between water and land use planning and management is considered a critical part of IWRM. Climate change adds another complication to the water and land interactions. Land use change and land cover change may impact on climate due to impacts on sources and stocks of greenhouse gases. Changes in irrigation infrastructure and water use patterns in agriculture may have an impact on water availability and demand. Adaptation of land use planning and management to climate change could help increase the adaptive capacity of the water sector. It is therefore necessary to coordinate climate change adaptation between water and land sectors, for which IWRM may be able to provide a framework to support and facilitate it.

Integration of Multiple Values of Water Resources Where water scarcity increases, using water to its maximum possible efficiency is preferred. Traditionally, water values tend to be appreciated in the case of off-stream use, where water use is fulfilled by abstraction. Water's in-stream values, such as those for environmental, recreational, and cultural purposes, did not receive legal protection until relatively recently (one exception is navigation). These in-stream values include aesthetic enjoyment, cultural and spiritual needs, tourism, recreation and leisure, environmental and ecological conservation, and waste disposal. The traditional water governance framework is mainly based on consumptive use, which demonstrates inability to accommodate increasingly valued in-stream uses. Integrating environmental objectives of water use into water allocation and management is one of the key elements of the IWRM paradigm, by which sustainable water resources management is sought to be achieved.[142] Integration of new values of water resources has been increasingly reflected in a range of water allocation mechanisms and institutional arrangements including environmental flows, cultural flows, and property rights to in-stream flows. According to the International Union for Conservation of Nature (IUCN) guide "Flow: the Essentials of Environmental Flows", the concept of environmental flows refers to "the water regime provided within a river, wetland or coastal zone to maintain ecosystems and their benefits where there are competing water uses and where flows are regulated".[143] Cultural flows is a concept that emerged over the last decade or so in discussion of indigenous interests in water planning and governance in Australia. The concept promotes a holistic approach to cultural values of water by indigenous people for environmental, spiritual, social, and economic purposes.[144] In the arid west of the US, the doctrine of prior appropriation is the dominant law of water allocation, which originated in the custom of

---

[141] IPCC. *Climate Change and Water*, 8.

[142] Ibid, 47.

[143] Megan Dyson et al., eds. *Flow: The Essentials of Environmental Flows*, Gland, Switzerland: International Union for Conservation of Nature and Natural Resources (IUCN), 2003, 3.

[144] Jessica Weir, "Cultural Flows in Murray River Country," *Australian Humanities Review* 48 (2010): 119–129.

"first in time, first in right".[145] Ground on "beneficial use" of water (traditionally consumptive and off-stream uses), prior appropriation results in difficulty in accommodating new and valuable in-stream uses. Concerted efforts have been observed across the western US, to incorporate in-stream uses of water into the property right regimes by recognising in-stream uses as "beneficial use", and to explore how tradable water rights and water markets can enable environmental protection and other public values of water resources.[146] Adding to the impacts of population growth, changing economic activities, land use change, urbanisation, and changing lifestyle, adverse effects of climate change on water resources further challenge water governance that is expected to balance multiple values of water and to address continued competition among water users. A more integrated approach to water governance may help to accommodate the changing societal views regarding water values, to facilitate negotiations, and to resolve conflicts between competing water users.

Inclusive Stakeholder Participation One IWRM principle is that water development and management should be based on a participatory approach, involving water users, planners and policy makers at all levels.[147] Participation requires that stakeholders at all levels of the social structure take part in the decision-making process and have an impact on decisions at different levels of water management.[148] In order for meaningful participation to take place, not only participatory mechanisms need to be established, but also the economic resources and information access, especially for disadvantaged groups, are necessary to facilitate their participation.[149] It is also essential to enhance participatory capacity of the disadvantaged groups so that they are able to effectively represent their interests in the decision-making process.[150] This principle finds resonance with adaptation to climate change, which similarly calls for a participatory approach to the adaptation process, from vulnerability assessment, evaluation of adaptation options, to implementation of adaptation strategies. Inclusive stakeholder participation helps coordinate interests, which in the water domain can be highly competing and conflicting. Expected to serve as an instrument to explore adaptation measures to climate change, IWRM, through effective stakeholder participation, may be able to capture society's views and

[145] Nicole L. Johnson, "Property without Possession," *Yale Journal on Regulation* 24 (2007): 205–251.

[146] Charlton H. Bonham, "Perspectives from the Field: A Review of Western Instream Flow Issues and Recommendations for a New Water Future," *Lewis & Clark Law School Environmental Law* 36 (2006): 1205–1235, 1223. Also see Charles W. Howe, "Protecting Public Values in a Water Market Setting: Improving Water Markets to Increase Economic Efficiency and Equity," *University of Denver Water Law Review* 3 (2000): 357–372, 370.

[147] Global Water Partnership Technical Advisory Committee, *Integrated Water Resources Management*.

[148] Cap-Net, *IWRM as a Tool for Adaptation to Climate Change: Training Manual and Facilitator's Guide*, South Africa: Cap-Net UNDP, 2009.

[149] Global Water Partnership Technical Advisory Committee, *Integrated Water Resources Management*.

[150] Cap-Net, *IWRM as a Tool for Adaptation to Climate Change*.

reshape planning processes of water resources, reducing the vulnerability and increasing the resilience of the water sector in turn.

2.3 Concluding Remarks

Water links multiple users, including humans and the environment, with each other in its integral hydrological cycle. The mobile and fugitive nature of water complicates the interdependency of coexisting water users. In the context of various existing stressors (economic development, growing population, lifestyle changes, and changing values of ecosystem services) coupled with climate change, how water governance responds to those challenges is at the centre of dealing with water scarcity, variability and uncertainty. After a brief overview of China's water scarcity and the key characteristics of water as a resource, this chapter provides an introduction to the recent developments in water governance around the implementation of market-oriented allocation mechanisms. It covers a number of key concepts which have significant implications for the legal and institutional arrangements of tradable water rights.

The unique characteristics of water resources and the complexity of water use have signification implications for shaping water governance. As water scarcity worsens is many places around the world, water governance is increasingly recognised as a central issue of water resource challenges. Water allocation mechanisms deal with multiple values of water and set the rules for water use, lying at the heart of water governance. International water reform debates around water as a social and economic good, water property regimes, institutional alternatives for water allocation, IWRM and climate change adaptation, all relate to the issue of how to allocate water in an era of constrained water availability. Water problems tend to be localised in a country or local area, contributing to shaping water governance suitable for the specific context. Diversity in the system of water rights trading is observed across jurisdictions. Evolution in water rights experiences adjustments and movements between different institutional arrangements of water allocation, highlighting the wrestle between government and market. As a global paradigm, IWRM attracts increasing attention from both policy and research for its potential in adaptation to climate change, and yet its implementation in practice remains a big challenge.

Against this backdrop, the research presented in this book aims to contribute to the international water governance debate by exploring from a legal and policy perspective, how China, comparative to other cases around the world, can find a balanced combination of water allocation mechanisms to address its water challenges. Well aware that effectiveness of water markets and tradable water rights in sustainable water use remains to be a continuing debate,[151] this research accepts the direction towards tradable water rights a given reality of China's water reform. It does

[151] For example, see Easter and Huang, *Water Markets for the 21st Century*.

not necessarily base its argument on market as the only or better solution for water allocation. Rather, it accepts the fact that China is now moving from an extreme government-based water allocation mechanism towards increased application of market instruments. As a result, the book intends to examine how law and policy can facilitate such transformation, and help China's water reform find the balance between government intervention and market freedom. The next chapters (Chaps. 3 and 4) will review the pathway of China's water law and policy reform as well as water trading practices, across different water allocation instruments including technological, administrative, and other economic instruments. In so doing, it hopes to provide a starting point for further review of the existing legal and policy system of water rights trading.

References

ACIL Tasman and Freehills. (2004). *An effective system of defining water property titles, Research report to the Australian government department of agriculture, fisheries and forestry, and land & water Australia*. Canberra: Land & Water Australia.

Anderson, T. L. (1983). The water crisis and the new resources economics. In T. L. Anderson (Ed.), *Water rights: Scarce resource allocation, bureaucracy, and the environment*. San Francisco/Cambridge, MA: Pacific Institute for Public Policy Research/ Ballinger Pub..

Anderson, T. L., Scarborough, B., & Watson, L. R. (2012). *Tapping water markets*. New York: Routledge.

Anonymous. (2007). Water prices for the priceless?: Implementing the justifiability of the right to water. *Harvard Law Review, 120*, 1076–1078.

Bartley, M., Fisher, D., & Queensland University of Technology. (2004). *Trading in water rights: Towards a national legal framework*. Sydney/Brisbane/Phillips Fox: Queensland University of Technology.

Bauer, C. J. (1998). *Against the current: Privatization, water markets, and the state in Chile*. Boston/London: Kluwer Academic.

Bauer, C. J. (2004). *Siren song: Chilean water law as a model for international reform*. Washington, DC/Great Britain: Resources for the Future.

Bauer, C. J. (2015). Water conflicts and entrenched governance problems in Chile's market model. *Water Alternatives, 8*(2), 147–172.

Bluemel, E. B. (2004). The implications of formulating a human right to water. *Ecology Law Quarterly, 31*, 957–1006.

Bonham, C. H. (2006). Perspectives from the field: A review of western instream flow issues and recommendations for a new water future. *Lewis & Clark Law School Environmental Law, 36*, 1205–1235.

Bonyhady, T., Macintosh, A., & McDonald, J. (Ed.). (2010). *Adaptation to climate change: Law and policy*. Sydney: The Federation Press.

Bromley, D. W. (1991). *Environment and economy: Property rights and public policy*. Oxford/Cambridge: Blackwell.

Bruns, B. R., & Meinzen-Dick, R. (2005). Frameworks for water rights: An overview of institutional options. In R. B. Bruns, C. Ringler, & R. Meinzen-Dick (Ed.), *Water rights reform: Lessons for institutional design* (pp. 3–25). Washington, DC: International Food Policy Research Institute.

Cap-Net. (2009). *IWRM as a tool for adaptation to climate change: Training manual and facilitator's guide*. Pretoria: Cap-Net UNDP.

Chen, J., & Xingmin, M. (2000). Tendency, causes and control measures on yellow river dry-up. *Journal of Natural Resources, 15*(1), 31–35.

China Institute of Geo-Environment Monitoring (CIGEM). (2014). *China groundwater resources*. http://219.142.70.139/dxs/HydRes.htm#huanjing. Accessed 9 July 2014.

China Ministry of Environmental Protection (MEP). (2013). *Report on the state of the environment in China*. 2014. http://www.zhb.gov.cn/zhxx/hjyw/201406/W020140605385940287254.pdf. Accessed 24 June 2014.

China Ministry of Water Resources (MWR). (2004). *Ministry of water resources 2003 annual report*. Beijing: Ministry of Water Resources.

China Ministry of Water Resources (MWR) and National Bureau of Statistics (NBS). (2012). *Bulletin of first national census for water*. Beijing: China Water and Power Press.

China National Bureau of Statistics (NBS). (2014). *China 2013 statistical yearbook*. http://www.stats.gov.cn/tjsj/ndsj/2013/indexch.htm. Accessed 24 Oct 2016.

Cruz, R. V., Harasawa, H., Lai, M., & Shaohong, W. (2007). Asia. In P. Martin, O. F. Canziani, J. P. Palutikof, P. J. van der Linden, & C. E. Hanson (Ed.), *Climate change 2007: Impacts, adaptation and vulnerability. Contribution of working group II to the fourth assessment report of the intergovernmental panel on climate change* (pp. 469–506). Cambridge: Cambridge University Press.

Dinar, A., Rosegrant, M. W., & Meinzen-Dick, R. (1997). *Water allocation mechanisms: Principles and examples, World Bank policy research working paper, no. 1779*. Washington, DC: World Bank.

Downs, P. W., Gregory, K. J., & Brookes, A. (1991). How integrated is river basin management? *Environmental Management, 15*(3), 299–309.

Dyson, M., Bergkamp, G., & Scanlon, J. (Ed.). (2003). *Flow: The essentials of environmental flows*. Gland: International Union for Conservation of Nature and Natural Resources (IUCN).

Easter, K. W., & Huang, Q. (Ed.). (2014). *Water markets for the 21st century: What have we learned?* Dordrecht: Springer.

Epstein, R. A. (2005). The historical variation in water rights. In J. Bennett (Ed.), *The evolution of markets for water: Theory and practice in Australia* (pp. 24–37). Cheltenham/Northampton: Edward Elgar.

Feng, W., Zhong, M., Lemoine, J.-M., Biancale, R., Hsu, H.-T., & Xia, J. (2013). Evaluation of groundwater depletion in north china using the gravity recovery and climate experiment (GRACE) data and ground-based measurements. *Water Resources Research, 49*, 2110–2118.

Gain, A. K., Rouillard, J. J., & Benson, D. (2013). Can integrated water resources management increase adaptive capacity to climate change adaptation? A critical review. *Journal of Water Resource and Protection, 5*, 11–20.

Global Water Partnership Technical Advisory Committee. (1998). *Integrated water resources management*. TAC background papers, No. 4.

Hill, M. (2012). *Climate change and water governance: Adaptive capacity in Chile and Switzerland*. Dordrecht/Heidelberg/New York/London: Springer.

Holden, P., & Thobani, M. (1996). *Tradable water righs: A property right approach to resolving water shortages and promoting investment*. World Bank policy research working paper 1627.

Howe, C. W. (2000). Protecting public values in a water market setting: Improving water markets to increase economic efficiency and equity. *University of Denver Water Law Review, 3*, 357–372.

Hu, D. (2006). *Water rights: An international and comparative study*. London: IWA Publishing.

Intergovernmental Panel on Climate Change (IPCC). (2013). Summary for policymakers. In T. F. Stocker, D. Qin, G.-K. Plattner, M. Tignor, S. K. Allen, J. Boschung, A. Nauels, Y. Xia, V. Bex, & P. M. Midgley (Ed.), *Climate change 2013: The physical science basis. Contribution of working group I to the fifth assessment report of the intergovernmental panel on climate change*. Cambridge/New York: Cambridge University Press.

IPCC. (2008). *Climate change and water, IPCC Technical Report VI*. IPCC: Geneva.

IPCC. (2011). *Special report on managing the risks of extreme events and disasters to advance climate change adaptation (SREX), summary for policy makers*. IPCC: Geneva.
Johnson, N. L. (2007). Property without possession. *Yale Journal on Regulation, 24*, 205–251.
Ludwig, F., van Slobbe, E., & Cofino, W. (2014). Climate change adaptation and integrated water resources management in the water sector. *Journal of Hydrology, 518*(Part B), 235–242.
Milly, P. C. D., Betancourt, J., Falkenmark, M., Hirsch, R. M., Kundzewicz, Z. W., Lettenmaier, D. P., & Stouffer, R. J. (2008). Stationarity is dead: Whither water management? *Science, 319*, 573–574.
Olmstead, S. M. (2014). Climate change adaptation and water resource management: A review of the literature. *Engergy Economics, 46*, 500–509.
Organisation for Economic Co-operation and Development (OECD) and Ministry of Agriculture of China (MOA). (2006). *Environment, water resources and agricultural policies: Lessons from China and OECD countries*. Paris: OECD.
Ostrom, E. (1990). *Governing the commons, the evolution of institutions for collective action*. Cambridge/New York: Cambridge University Press.
Ostrom, E. (2000). Private and common property. *Encyclopedia of Law and Economics, II*, 332–379.
Perry, C. J., Rock, M., & Seckler, D. (1997). *Water as an Economic Good: A Solution, or a Problem? International Irrigation Management Institute Research Paper, 14*. Colombo: International Irrigation Management Institute.
Qiu, J. (2010). China faces up to groundwater crisis. *Nature, 466*, 308.
Rouillard, J. J., Benson, D., & Gain, A. K. (2014). Evaluating IWRM implementation success: Are water policies in Bangladesh enhancing adaptive capacity to climate change impact? *International Journal of Water Resources Development, 30*(3), 515–527.
Sadeque, S. Z. (2000). Nature's bounty or scarce commodity: Competition and consensus over groundwater use in rural Bangladesh. In B. R. Bruns & R. Meinzen-Dick (Ed.), *Negotiating water rights* (pp. 269–291). London: ITDG Publishing.
Salzman, J. (2005). *Creating markets for ecosystem services: Notes from the field*. Duke Law School Science, technology, and innovation research paper no. 2.
Schlager, E., & Ostrom, E. (1992). Property Rights regimes and natural resources: A conceptual analysis. *Land Economics, 68*(3), 250–251.
Scott, A., & Coustalin, G. (1995). The evolution of water rights. *Natural Resources Journal, 35*(4), 821–979.
Segerfeldt, F. (2005). *Water for sale: How business and the market can resolve the world's water crisis*. Cato Institute: Washington, DC.
Sultana, F., & Loftus, A. (Ed.). (2012). *The right to water: Politics, governance and social struggles*. London/New York: Earthscan.
UNDP. (1997a). *Governance for sustainable human development. A UNDP policy document*. New York: UNDP Management Development and Governance Division/Bureau for Policy and Programme Support.
UNDP. (1997b). *Reconceptualising governance, Discussion paper 2*. New York: UNDP Management Development and Governance Division/Bureau for Policy and Programme Support.
United Nations Development Programme (UNDP). (2014). *Discussion paper: Governance for sustainable development: Integrating governance in the post-2015 development framework*. New York: UNDP.
Watson, N. (2004). Integrated river basin management: A case for collaboration. *International Journal of River Basin Management, 2*(4), 243–257.
Weir, J. (2010). Cultural flows in murray river country. *Australian Humanities Review, 48*, 119–129.
World Bank. (2014). *World development indicators: Table 3.5: Freshwater*. http://wdi.worldbank.org/table/3.5. Accessed 24 Oct 2016.
World Health Organisation (WHO) and United Nations Children's Fund (UNICEF). (2014). *Progress on drinking water and sanitation (2014 update)*. Geneva: WHO/UNICEF.

World Water Assessment Programme (WWAP). (2014). *Water and energy.* The united nations world water development report 5. http://www.unesco.org/new/en/natural-sciences/environment/water/wwap/wwdr/2014-water-and-energy/. Accessed 24 Oct 2016.

WWAP. (2003). *Water for people, water for life, The united nations world water development report 1.* Paris: UNESCO/Berghahn Books. http://www.unesco.org/new/en/natural-sciences/environment/water/wwap/wwdr/. Accessed 17 Nov 2014.

WWAP. (2006). *Water, a shared responsibility, the united nations world water development report 2.* Paris: UNESCO. http://www.unesco.org/new/en/natural-sciences/environment/water/wwap/wwdr/. Accessed 17 Nov 2014.

WWAP. (2009). *Water in a changing world, The united nations world water development report 3.* Paris: UNESCO/Earthscan. http://www.unesco.org/new/en/natural-sciences/environment/water/wwap/wwdr/. Accessed 17 Nov 2014.

WWAP. (2012). *Managing water under uncertainty and risk, The united nations world water development report 4.* Paris: UNESCO. http://www.unesco.org/new/en/natural-sciences/environment/water/wwap/wwdr/. Accessed 17 Nov 2014.

Yang, H., Zhang, Z., & Shi, M. (2012). The impact of China's economic growth on its water resources: A regional and sectoral assessment. In H. McKay & L. Song (Ed.), *Rebalancing and sustaining growth in China* (pp. 309–328). Canberra: ANU E Press.

Zhang, Y. (2014). *The deputy director of the state forestry administration's speech at the press conference for the findings of the second national wetland resources survey.* http://www.china.com.cn/zhibo/2014-01/13/content_31140091.htm?show+t. Accessed 9 July 2014.

Zhang, L., Xia, J., & Zhifang, H. (2009). Situation and problem analysis of water resources security in China. *Resources and Environment in the Yangtze Basin, 18*(2), 116–120.

Legal and Policy Sources

Domestic

The 11th Five Year Plan of Water Resources Development of China (China National Development and Reform Commission, Ministry of Water Resources, and Ministry of Construction). 2007. http://www.sdpc.gov.cn/zcfb/zcfbghwb/200706/W020140220601664956788.pdf. Accessed 24 Oct 2016.

The 12th Five Year Plan for Environmental Risk Control of Chemicals (China Ministry of Environmental Protection), 2013. http://www.mep.gov.cn/gkml/hbb/bwj/201302/W020130220539067366659.pdf. Accessed 8 July 2014.

The 12th Five Year Plan for Rural Drinking Water Safety Projects (China National Development and Reform Commission, Ministry of Water Resources, Ministry of Health, and Ministry of Environmental Protection), 2014. http://www.sdpc.gov.cn/zcfb/zcfbghwb/201402/P020140221360445500781.pdf. Accessed 8 July 2014.

The Water Resources Development Plan (2011-2015) (China National Development and Reform Commission, Ministry of Water Resources, and Ministry of Housing and Urban-Rual Development), 2012. http://www.sdpc.gov.cn/fzgggz/ncjj/nczc/201206/t20120621_487019.html. Accessed 24 June 2014.

International

United Nations General Assembly. (2010). *Resolution adopted by the general assembly on 28 July 2010*. A/RES/64/292. United Nations. http://contrattoacqua.it/public/upload/1/2/tab_elms_docs/1404320102n0947935.pdf. Accessed 2 May 2017.
United Nations Human Rights Council. *Resolution Adopted by the Human Rights Council, 15/9 Human Rights and Access to Safe Drinking Water and Sanitation*. A/HRC/RES/15/9. United Nations, 2010.
United Nations Economic and Social Council. *General Comment No. 15 on the Right to Water*. 2002. http://www2.ohchr.org/english/issues/water/docs/CESCR_GC_15.pdf. Accessed 2 May 2017.

Chapter 3
China's Water Law and Policy Reform: How Far Have We Travelled?

Abstract Dominated by the philosophy of "man can conquer nature", China's water governance was traditionally oriented towards construction and utilisation rather than conservation and protection of water resources. It failed to encourage water saving and high water use efficiency due to its ineffective administrative allocation mechanism and fragmented management. Reflecting on the failure of the traditional notion of water use, and in particular its perception on the human-nature relationship, China has been undergoing a large-scale water reform, "Building a water saving society". This water reform aims to achieve "harmonious coexistence between man and nature" by establishing a national water right system. This chapter reviews the current legal and policy developments in water rights trading in China. It provides an insightful understanding of the legal and institutional context within which tradable water rights have to be developed and operated. China is a country that has been experiencing fundamental changes in its economic, social, political, and legal systems over the past three decades. The evolution of China's approaches to water management is representative of such transition. This chapter therefore also reviews the institutional framework of China's water resources management, including laws, policies, and regulatory frameworks related to the key issues in water resources management. Such understanding is critical as it provides an overview of the starting point for future legal and policy reform towards tradable water rights.

Keywords China's water reform • Building a water saving society • Legal framework • Institutional framework • Water rights system

3.1 China's Legal Framework for Water Governance

China's water legal system began to develop in the 1980s, when utilisation of water resources to address water shortages started becoming an increasingly critical challenge to support social and economic development. China has established an integral portfolio of laws and policies to regulate water-related activities after more than three decades of development. This portfolio consists of a range of legal documents at multiple levels, including: (1) constitution by the National People's Congress (NPC); (2) legislations by the standing committee of the Congress; (3) regulations

M. Jiang, *Towards Tradable Water Rights*, Global Issues in Water Policy 18,
DOI 10.1007/978-3-319-67087-4_3

by the State Council; (4) provisions or rules by ministries or administrations under the State Council; and (5) regulations and implementing rules by the provincial and other local legislatures and governments.[1] There are four types of regulations: (1) regulations and other documents having the force of law issued by the State Council; (2) ministerial regulations, national standards and rules issued by state ministries or commissions; (3) regulations issued by people's congresses and their standing committees at the provincial and other local levels; and (4) regulations and other legal orders issued by provincial and other local governments.[2] The following figure (Fig. 3.1) shows China's legal framework highlighting the laws and regulations related to water management and tradable water rights.

The Constitution stands at the highest level in the legal hierarchy of China. It is the fundamental law of the nation and has supreme legal authority.[3] No law or administrative or local rules and regulations shall contravene the Constitution.[4] Article 9 of the 1982 Constitution of the People's Republic of China (2004 Amendment) establishes the state ownership of natural resources including water resources:

> (1) Mineral resources, waters, forests, mountains, grassland, unreclaimed land, beaches, and other natural resources are owned by the state, that is, by the whole people, with the exception of the forests, mountains, grassland, unreclaimed land, and beaches that are owned by collectives in accordance with the law.[5]

The Constitution's Article 26 then refers to environmental protection: "The state protects and improves the living environment and the ecological environment, and prevents and remedies pollution and other public hazards".[6] The Environmental Protection Law was formulated in 1989, based on the 1979 Trial Environmental Protection Law. Its Article 2 defines "the environment" as:

> The total body of all natural elements and artificially transformed natural elements affecting human existence and development, which includes the atmosphere, water, seas, land, minerals, forests, grassland, wildlife, natural and human remains, nature reserves, historic sites and scenic spots, and urban and rural areas.[7]

---

[1] Legislation here only refers to the laws enacted by the NPC and its standing committee of China. Unlike states in other foreign countries, China's provinces have no power to make legislations. Rather, they can only produce implementing rules for legislations or provincial regulations in their local context.

[2] William P. Alford and Yuanyuan Shen, "Limits of the Law in Addressing China's Environmental Dilemma," *Stanford Environment Law Journal* 16 (1997): 125–148. Also see Stefanie Beyer, "Environmental Law and Policy in the People's Republic of China," *Chinese Journal of International Law* 5, no. 1 (2006): 185–211.

[3] 1982 Constitution of the People's Republic of China (2004 Amendments), http://www.npc.gov.cn/englishnpc/Constitution/node_2825.htm (accessed May 21, 2017), Preamble, the last paragraph.

[4] Ibid, art. 5:3.

[5] Ibid, art. 9.

[6] Ibid, art. 26.

[7] 1989 Environmental Protection Law of the People's Republic of China, http://www.npc.gov.cn/englishnpc/Law/2007-12/12/content_1383917.htm (accessed May 21, 2017), art. 2.

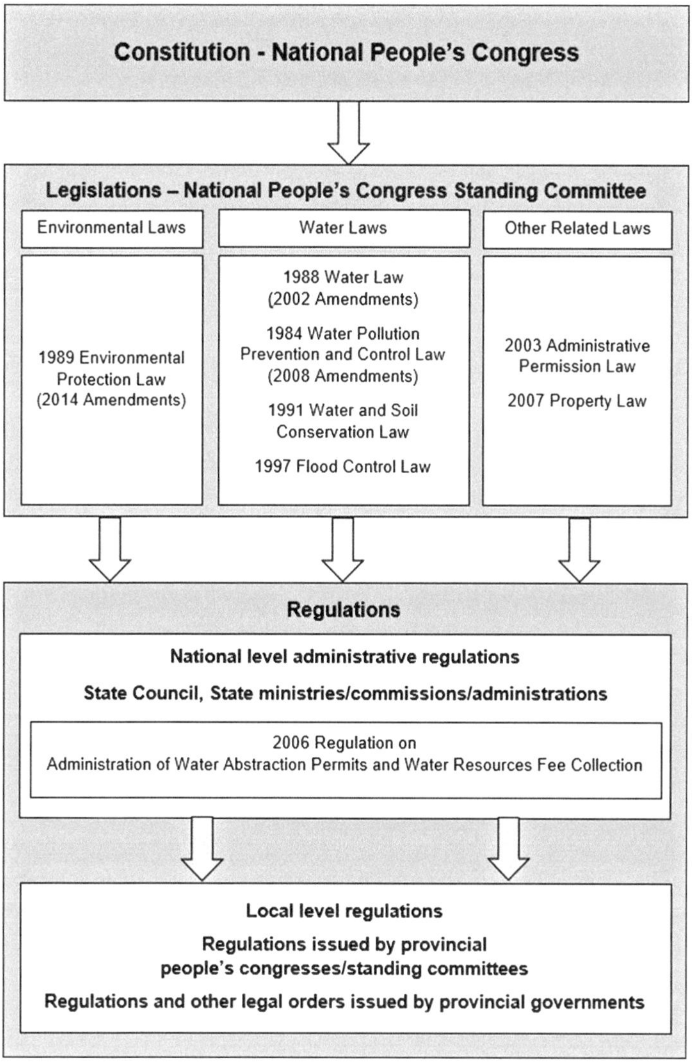

Fig 3.1 China's legal framework for water management and tradable water rights

The 1989 Environmental Protection Law has clearly demonstrated its limitations stemming from the planned economy tradition and failure to respond to the increasing pressures on the environment posed by the rapid economic growth. However, the amendment did not take place until April 2014, 25 years after the law had come into effect. With more updated and specified environmental protection goals and stiffer punishments, the revised law is expected to enhance the enforcement and compliance.

Aiming to address conflicts and shortfalls of the deficient and fragmented system that arose during the 1970s and early 1980s, the 1988 Water Law is the first fundamental legislation regulating water activities.[8] The provincial congresses and their standing committees adopted local implementing regulations of the water law in their specific local contexts. The water law has established an institutional framework for China's water resources administration, called the Competent Department of Water Administration (CDWA), which has played an important role in regulating water resources utilisation and development. However, the law failed to effectively address water shortages and environmental degradation in China's water resources. By encouraging utilisation of water resources rather than water saving, the law facilitated the booming of thousands of engineering projects around the country. As a result, in 2002, the 1988 Water Law was substantially amended in order to better cope with the increasing water demands and protect the severely polluted water bodies. Following this amendment, locally implemented water regulations at the provincial level were also modified in accordance with the new law.

Article 2 of the revised Water Law applies to the development, utilisation, conservation, protection, and management of water resources, including both surface and groundwaters.[9] The main provisions of the 2002 Water Law as opposed to the 1988 Water Law are highlighted in the following table (Table 3.1).

Despite the positive step the new Water Law took forward, it has been criticised for several reasons. Similar to other environmental laws in China, the new Water Law suffers from vagueness, being too general and exhortational, leading to difficulty in implementation.[10] Another important deficiency of the new Water Law is its failure to recognise the integrated water resources management (IWRM). Responding to the need for a more integrated approach to water management, the amendment has strengthened river basin management. Nevertheless, it made a big compromise to the existing political system, in which fragmentation in water management represents competing interests of different administrative agencies. The new Water Law also takes a conservative approach to the market-based mechanisms for water allocation. Although the pilot programmes of water trading have been successfully implemented in several regions, tradable water rights are left ill-defined in the amendment. Water trading is not explicitly allowed in the 2002 Water Law, which still views the administrative allocation mechanisms as the main instrument. Therefore, it missed the opportunity to facilitate the development of the market-based water allocation mechanisms.

There are three other laws directly related to water resources in the current legal framework: the 1984 Law on Prevention and Control of Water Pollution (2008 Amendment), the 1991 Water and Soil Conservation Law, and the 1997 Flood Control Law. Together with the 2002 Water Law, they form the core foundation of

---

[8] Patricia Wouters et al., "The New Development of Water Law in China," *University of Denver Water Law Review* 7 (2004): 259.

[9] 2002 Water Law of the People's Republic of China, http://www.npc.gov.cn/englishnpc/Law/2007-12/12/content_1383920.htm (accessed May 21 2017), art. 2.

[10] Elizabeth C. Economy, *The River Runs Black: The Environmental Challenge to China's Future* (Ithaca, N.Y.: Cornell University Press, 2004), 101. Also see Beyer, "Environmental Law and Policy in the People's Republic of China".

Table 3.1 Key amendments in China's Water Law (1988 vs. 2002)

Provision	1988 Water Law	2002 Water Law
Purpose of the law	Article 1: the full derivation of comprehensive benefits from water resources.	Article 1: the sustainable utilisation of water resources.
Ownership of water resources	Article 3: state property and common property.	Article 3: state property.
Water withdrawal rights	Rights to withdraw water were not explicitly acknowledge.	Article 48: the recognition of water use rights.
Water resources management	Article 9: combination of management based on national, local and departmental levels.	Article 12: combination of management based on national, basin and local levels. River basin management is strengthened, and river basin authorities are clearly mandated regulatory powers in water management, administration and supervision.
Water planning	Article 11: only one general statement.	Chapter 2 (Articles 14–19): the various aspects of water planning are specified including types of plans, makers, and approval procedures.
Water saving	Water saving was given little attention while encouraging water utilisation.	Article 8: 'building a water-saving society' as a fundamental principle of the law, calling for a technological, economic, legal and institutional revolution of water use in China.
Water protection	Water resources protection was given little attention.	New systems are established to protect water resources and environments.
		Article 21: environmental needs for water.
		Articles 32 and 33: water functional zones and protection of potable water sources.

China's water legal system. The Law on Prevention and Control of Water Pollution deals with water pollution and water environmental protection. As its Article 2 states, it applies to prevention and control of pollution of rivers, lakes, canals, irrigation channels, reservoirs and other surface water bodies and groundwater bodies within China's territory.[11] This law has established an institutional framework for water pollution control and water environments protection, which is different from the system for protection of water resources under the 2002 Water Law. Such man-made distinction between water resources and water environments complicates institutional arrangements and results in conflicts in implementation. In response to the continuous water deterioration of China's water bodies, it went onto the agenda of the NPC's Standing Committee in May 2014 for a new amendment.

Two laws outside the field of environmental law deserve to be highlighted here for their significance to water allocation in China. The first one is the 2003

[11] 1984 Law of the People's Republic of China on Prevention and Control of Water Pollution Prevention (2008 Amendments), http://www.mwr.gov.cn/english/laws.html (accessed May 22, 2017), art. 2.

Administrative Permission Law. Provisions about administrative licences in this law include its establishment, regulatory organisations, procedures, expenses, and supervision. Its Article 9 generally prohibits transfer of administrative licences by stating that

> An administrative license obtained in accordance with the law shall not be transferred except for those that may be transferred in light of the legal conditions and procedures of the laws and regulations.[12]

Water abstraction permits, as a type of administrative licence, are governed by this law. Accordingly, water abstraction permits are not transferrable unless laws and regulations otherwise provide specific exceptions.

The other legislation noteworthy is the 2007 Property Law. The state ownership of water resources is reaffirmed in Article 46 of this law. Based on this state ownership of natural resources including water, the usufructuary rights to use the natural resources are established in Article 118:

> As regards the natural resources that are owned by the state or that are owned by the state but used by the collectives as well as those that are owned by the collectives as prescribed by law, an entity or individual may possess, use and seek proceeds from them.[13]

The 2007 Property Law also states that rights related to use of water resources, such as water abstraction rights, the rights to use water for breeding or fishery are under the protection of law.[14]

To implement the provisions of these water-related laws, regulations and rules are established by the central government (the State Council), its ministries and administrations, and legislatures at the local level. These regulations cover various issues of water resources management including water administration, water and soil conservation, water quality and pollution control, flood control, water projects (dams and reservoirs) management, river course management, water abstraction permits, water saving, river basin management, water supply and pricing, and irrigation management. Generally speaking, China has established a fairly comprehensive body of water regulations.

In order to understand the legal context of water rights and water rights administration in China, it is necessary to highlight the 2006 Regulation on Administration of Water Abstraction Permits and Water Resources Fee Collection. This regulation replaces the 1993 Implementing Measures on Water Abstraction Permits System and provides a new framework for granting and managing water abstraction permits.[15] Most importantly, unlike the old implementing measures that prohibited any

---

[12] 2003 Administrative Permission Law of the People's Republic of China, http://www.npc.gov.cn/englishnpc/Law/2007-12/05/content_1381966.htm (accessed May 22, 2017), art. 9.

[13] 2007 Property Law of the People's Republic of China, http://www.npc.gov.cn/englishnpc/Law/2009-02/20/content_1471118.htm (accessed May 21, 2017), art. 118.

[14] Ibid, art. 123.

[15] Australia China Environment Development Program, *Report on the Development of Water Rights and Trading in the People's Republic of China*, Canberra: Department of Agriculture, Fisheries and Forestry of Australia, 2006, 8.

form of transfer of water abstraction permits,[16] this new regulation loosened the restriction by allowing transfers under certain circumstances. As its Article 27 states,

> Entities or individuals who have obtained water abstraction rights in accordance with law, with the approval from the original permitting authority, may transfer the water resources that they have conserved by the adjustment in product and industrial structures, technology innovation or other water saving measures for profit in accordance with law within the duration and quantity limits of water abstraction permits, and shall register the change of the water abstraction rights with the permitting authority. The competent department of water administration under the State Council shall formulate the detailed measures of the transfer.[17]

In response to pilot programmes of water rights trading in some local areas, this provision forms a major legal basis for water rights trading.

3.2 Institutional Framework of Water Governance in China

3.2.1 Regulatory Organisations of Water Governance

As briefly mentioned earlier, Water Law and Water Pollution Prevention and Control Law have established different institutional frameworks across which water quantity and quality issues are separately managed. Although the amended Water Law has abandoned the sectoral management of water resources, its compromise in IWRM ended up with the coexisting systems of river basin management and administrative unit-based management.[18] The administrative unit-based management system further combines administration at different levels and by different agencies. In the context of such combined management at national, basin, and local levels, the institutions involved in water resources management in China consist of a complicated hierarchy.

3.2.1.1 The Vertical Hierarchy

Within the system of the Competent Department of Water Administration (CDWA), the Ministry of Water Resources (MWR) is at the top level of the vertical hierarch, directly under the State Council. It is the highest water authority of China in charge of the unified administration and supervision of water resources of the country.

[16] 1993 Implementing Measures on the Water Abstraction Permits System (China State Council), http://www.chinawater.net.cn/law/W03.htm (accessed 28 September, 2008), art. 26.

[17] 2006 Regulation on Administration of Water Abstraction Permits and Water Resources Fee Collection (China State Council), http://www.gov.cn/zwgk/2006-03/06/content_220023.htm (accessed May 21, 2017), art. 27.

[18] China 2002 Water Law, art. 12.

Under the water law, the MWR's main mandates include formulation of water-related regulations and policies, implementation of water-related legislation, integrated management of water resources, and guidance to water-related activities. In terms of specific issues in water resources management, the MWR is responsible for management water resources development, utilisation, saving, conservation, flood prevention, and water allocation and planning.[19]

At the river basin level, the MWR has set up river basin authorities as its agencies for important rivers and lakes designated by the State Council, to exercise water administration and supervision in their respective boundaries. Their responsibilities can be either stipulated by laws and regulations or authorised by the MWR. There are seven major river/lake basins for which authorities are established[20]: the Changjiang (Yangtze River) Water Conservancy Commission, the Yellow River Conservancy Commission (YRCC), the Huai River Water Conservancy Commission, the Hai River Water Conservancy Commission, the Pearl River Water Conservancy Commission, the Song-Liao River Water Conservancy Commission, and the Taihu Basin Authority. Other basin authorities may also be established for better coordination among provinces concerned in trans-provincial rivers and lakes. River basin authorities are responsible for planning, enforcement, and monitoring water-related activities within their basin.[21]

The local governments, at the provincial, prefecture, and county levels, also play an important role in water resources administration. They are widely involved in water utilisation, conservation and protection, water planning, water disaster management, and water disputes resolution. Local governments above or at the county level have their CDWAs, namely, water resources bureaus (WRBs), or increasingly, more integrated water affairs bureaus (WABs).[22] As the counterparts of the MWR at the local levels, WRBs or WABs have responsibility for water management within their corresponding administrative areas. While these local water authorities may share water-related responsibilities with other departments, they may also take extended responsibilities in water supply, flood control and management, urban water saving, and wastewater treatment.[23]

The dual system of water management based on river basins and administrative units has demonstrated its failure in integrated management. In reality, trans-provincial river basins in China tend to be fragmentally managed due to weak river basin authorities and ineffective coordination among upstream and downstream provinces within the same basin. Not surprisingly, provincial governments compete against each other for water availability, and disputes take place frequently in these trans-provincial river basins, which river basin authorities are not strong enough to play their role in conflict resolution. One major reason resulting in this situation is

[19] Feng Sun, Zhongshan Yang, and Zhenfang Huang, "Challenges and Solutions of Urban Hydrology in Beijing," *Water Resources Management* 28 (2014): 3377–3389.

[20] Ibid.

[21] Ibid.

[22] Ibid.

[23] Ibid.

the hierarchical status of river basin authorities. As the MWR's agencies, river basin authorities stand one to three notches lower than provincial governments in China's bureaucratic hierarchy. As a result, river basin authorities have weakened power to successfully coordinate water allocation and resolve water disputes among provinces. Furthermore, while nominally responsible to both the MWR and their local governments, WRBs or WABs rely on the latter much more than the former as the local governments appoint their CDWAs' personnel and provide their financial budget. It is not surprising to see that the MWR maintains only a supervisory role to local WRBs or WABs, who tend to favour their local government's concerns over the MWR's national priorities in their decision making.

3.2.1.2 The Horizontal Hierarchy

Beside the vertical water management system, a horizontal administration system exists in China for the management of water resources. Such fragmentation in water management by various governmental departments therefore represents another institutional problem, expressed with an enduring complaint especially from the MWR that "too many dragons manage the water".[24] Depending on different issues related to water resources, water management duties are shared across many authorities as a result of the old Water Law and other water-related laws influenced by the traditional planned economy. Although the new Water Law has abolished the principle of departmental water management, other water-related laws are still in effect, leading to continuing fragmentation in water management.

Apart from the fact that the MWR acts as the core national water administration, many other ministries, commissions or administrations under the State Council take direct or indirect water-related responsibilities.[25] While the MWR is involved in almost all water resources management and takes the dominant mandates in water allocation, water abstraction permits, water and soil conservation, flood control and water projects, a range of other authorities are directly involved in water management. These authorities include the Ministry of Environmental Protection (MEP), the Ministry of Housing and Urban-Rural Development (MOHURD), the Ministry of Land and Resources (MLR), the National Development and Reform Commission (NDRC), the National Forestry Administration (NFA), and the National Health and Family Planning Commission (NHFPC). The following figure (Fig. 3.2) illustrates the responsibilities of water management in various water-related activities shared among these governmental agencies. If taking into account indirect water

[24] James E. Nickum and Yok-Shiu F. Lee, "Same Longitude, Different Latitudes: Institutional Change in Rrban Water in China, North and South," in *Environmental Governance in China*, ed. Neil T. Carter and Arthur P.J. Mol (Hoboken: Taylor and Francis, 2013), 84.

[25] Min Jiang, Dan Cao, and Terry DeLacy, "Conflict and Coordination of Multiple Values of Water Resources: A Legal and Policy Analysis from a Perspective of Fresh Water Based Tourism in China," In *Conference Proceeding of the International Conference on Drinking Water Safety, Security, and Sustainability*, ed. Miaomiao Ye, David Z. Zhu, and Tuqiao Zhang (Hangzhou: Zhejiang University, 2011), 134.

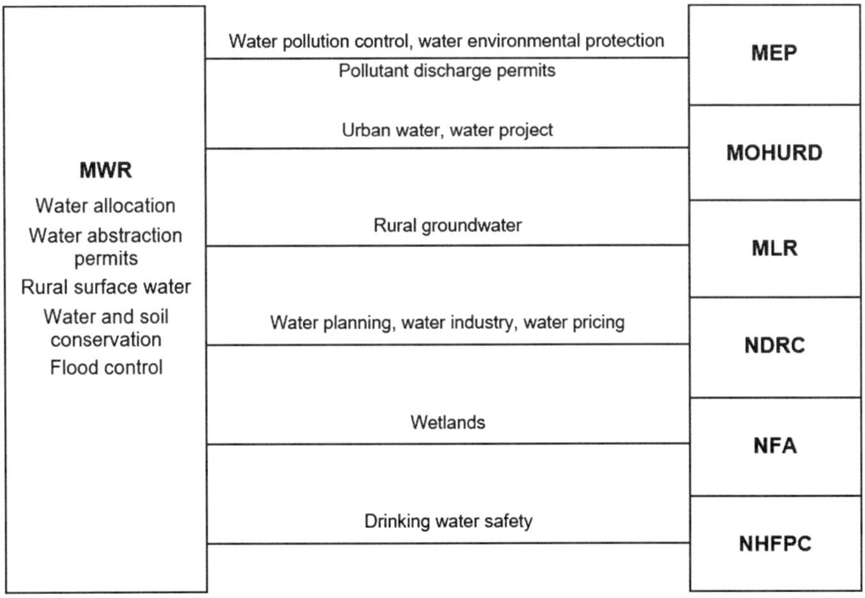

Fig 3.2 Water management responsibilities shared among national level administrations in China (Modified from Jiang, Cao, and DeLacy, 2011, 134)

management, the list extends to ministries or administrations of finance, agriculture, fishery, and transportation.

As a result of fragmented sector-oriented management of water resources, conflicts often arise between related authorities due to overlapping responsibilities and lack of effective coordination. The competitive relationship between the MWR and the MEP offers a typical example. As shown in the following table (Table 3.2), the MWR and the MEP, as well as their respective counterparts at local levels, WRBs or WABs and Environmental Protection Bureaus (EPBs), both have statutory responsibilities for water resources protection and pollution control based on the Water Law and the Water Pollution Prevention and Control Law.

Sharing the management roles in water protection and pollution control, the MWR and the MEP tend to be more competitive with each other than cooperative.[26] Indeed, the duplicated responsibilities not only compromise efficiency, but also result in inconsistencies, due to generally poor inter-agency communication. The MWR and the MEP maintain parallel, in many cases, overlapping national water quality monitoring systems, with different monitoring standards and techniques.[27] As a result, water quality data they publish show different evaluations of water quality, based on which they formulate plans with different requirements of waste discharge. The MWR's monitoring system takes both quantitative and qualitative

[26] World Bank, *China: Air, Land and Water: Environmental Priorities for a New Millennium*, Washington, D.C.: World Bank, 2001, 66.

[27] Ibid.

Table 3.2 Respective management functions of MWR and MEP in water resources protection

Water issues	MWR	MEP
Planning	Comprehensive plans and resource protection plans	Pollution control plans
Quality monitoring	Water resources quality	Water environment quality
Pollution control	Pollutant-carrying capacity of rivers and the total pollutant discharge limit	Pollutant discharge permits
	Direct pollutant discharge mouths to rivers and lakes	Pollution sources
		Wastewater treatment
Function zones	Water function zones	Water environment function zones

aspects of water bodies into account, which normally lead to stricter requirements than those of the MEP. Unfortunately, such requirements are very difficult for implementation because local EPBs would follow the MEP's standards in granting discharge permits. In addition, rivers and lakes are often divided into two different types of function zones: function zones by the MWR's categories for water use type, and function zones by the MEP's categories for water environment protection. The operation and management of the two sets of function zones in parallel causes waste of government resources and more importantly, lack of clarity confuses water users and other stakeholders, which hinders implementation and enforcement.

At the local levels, water management can be even more fragmented. Water management functions are divided and mandated to multiple departments, which are in charge of managing different water-related activities along with WRBs or WABs at provincial, prefecture and county levels. Accordingly, surface water and groundwater, water quality and water quantity, urban water and rural water, water supply and drainage are separately managed by different authorities. WRBs in some areas as the local water authorities, only take responsibilities of managing rural water, surface water, water projects, water and soil conservation, and flood control. Other authorities may come in charge of urban water management (departments of construction and urban planning), rural groundwater (departments of geology and mining), and drainage and pollution control (the departments of environmental protection). The sectoral fragmentation tends to create conflicts in water management as the statutory responsibilities shared among the authorities are often overlapping and difficult to clearly define. In response to this issue, the MWR has initiated a water institutional reform intending to promote a more integrated approach to water management. WRBs are transformed into WABs by obtaining broader water-related functions from other departments. By the end of 2012, more than 1900 local governments at and above the county level have either established WABs or extended their WRBs for a more integrated water management system, accounting for over 78% of the total local governments at and above the county level.[28] Despite the efforts in moving away from the former sector-bound approach

[28] China Ministry of Water Resources (MWR), *2012 Statistic Bulletin on China Water Activities*, Beijing: China Water and Power Press, 2012.

to water management, to what extent the new WABs system is able to serve better to integrate water management functions remains arguable. Critics question that what have been changed are just the names in most cases due to lack of capacity of WABs in managing the newly appointed responsibilities. In addition, the merge of departments and their mandates may only internalise conflict, instead of solving it.[29]

To summarise, there is generally a lack of both horizontal and vertical coordination at the institutional level of water resources management in China. Sectoral fragmentation remains prominent although efforts have been made towards a more integrated approach. Water management based on administrative units stands strong against river basin management as river basin authorities need to have a more strengthened position in the bureaucratic hierarchy. The CDWAs at local levels are also yet to build their capacity in taking expanded responsibilities in water management and work with other related departments in a more coordinated manner.

3.2.2 Environmental Non-Governmental Organisations (NGOs)

Besides a wide range of governmental agencies vertically and horizontally forming the centre of the institutional framework of China's water resources management, environmental non-governmental organisations (NGOs) have been developing at a tremendous speed in China, and increasingly playing a significant role in environmental protection.[30] Since the launch of the first environmental NGO, Friends of Nature[31] in 1994,[32] there were 3539 environmental civil society organisations (CSOs) in China (including Hong Kong, Macau, and Taiwan) by the end of October 2008.[33] These CSOs based in mainland China include government-organised NGOs (1309), voluntary grass-roots NGOs (508), student unions and associations (1382), and branches of international NGOs (90).[34]

Chinese environmental NGOs have been increasingly strengthening their influence on environmental protection, and particularly water resources conservation by

---

[29] Nickum and Lee, "Same Longitude, Different Latitudes".

[30] Robert V. Percival and Huiyu Zhao, "The Role of Civil Society in Environmental Governance in the United States and China," *Duke Environmental Law & Policy Forum* 24 (2014): 142.

[31] China Environmental Science Association is the first government-organised environmental NGO in China, set up in 1978.

[32] Percival and Zhao, "The Role of Civil Society in Environmental Governance in the United States and China".

[33] All-China Environment Federation, *Summary Report on the Development of China's Environmental Civil Society Organisations in 2008*, Beijing: All-China Environment Federation, 2008, 1.

[34] Ibid.

raising public awareness, observing and disclosing business activities and government failures, and participating in legal consultation. In 2006, the Public and Environment Research Institute released the first online water pollution database, namely China Water Pollution Map, which aims to disclose water quality facts and monitor polluters' performances in different areas of China.[35] As the institute's director, Jun Ma, who is also the author of China's Water Crisis, said, public participation in environmental management is not possible unless information becomes accessible to them.[36]

Objection to dam construction on the Nujiang River has been one of the biggest victories of environmental NGOs in influencing decision making in water projects. In 2003, three major environmental NGOs, namely Green River, Green Earth Volunteers, and Friends of Nature, jointly led a campaign to halt the construction of two reservoirs and 13 cascade hydroelectric power stations on the lower reaches of the Nujiang River in Yunnan Province, arguing that the environmental impacts of the proposal should be carefully assessed before the decision is made.[37] They also petitioned to the central government for more attention to dam construction. In response to it, the MEP organised a research expert team across disciplines of ecology, agriculture, forestry, geography, heritage protection, hydraulics, and biology to undertake a comprehensive assessment. The then Premier Wen Jiabao commented on the proposal that thorough investigation should be undertaken for those major hydropower projects of great concern to the society with major environmental implications.[38] The efforts of the campaigning environmental NGOs stimulated a heated debate on the issue of exploitation of the Nujiang River. As a result, the MEP committed that major revisions and adjustments shall be made for the original proposal to be approved.[39]

Another contribution of environmental NGOs in China is providing legal aid with environmental pollution victims, although this type of activity remains limited in scope and scale.[40] An exceptional example is the Centre for Legal Assistance to Pollution Victims (CLAPV) based in Beijing. The CLAPV started offering free legal assistance to pollution victims through its telephone hotline in 1999. Over eight years up to 2007, it has logged more than 10,000 calls and taken over 100

---

[35] Ibid.

[36] People's Net, *Water Pollution Map Made by Environmental NGOs to Disclosing Polluters*, People's Net, 2006, http://www.022net.com/2006/9-14/465750243044915.html (accessed September 25, 2008).

[37] Tianyu Li, *Environmental NGOs: Transition from Supporting Role to Leading Role*, *Journalist Observation*, 2007, http://hxtc.china.cn/environment/2007-07/02/content_8467000.htm (accessed September 25, 2008).

[38] Xinhua Net, *The Premier Wen Jiabao Gave Instructions to Speed up the Assessment of the Nujiang River Project*, 2005, http://www.china5e.com/news/water/200509/200509120049.html (accessed September 25, 2008).

[39] Lin Zuo, "Protecting Nujiang River: No Winners Yet," *New Century Weekly*, 2006, http://www.china5e.com/news/huanbao/200606/200606200051.html (accessed September 25, 2008).

[40] Xueyong Zhan and Shuiyan Tang, "Political Opportunities, Resources Constraints and Policy Advocacy of Environmental NGOs in China," *Public Administration* 91, no.2 (2013): 381–399.

cases to court, some with as many as 1700 plaintiffs.[41] Running the centre to "help people whose lives and livelihoods have been damaged by the toxic byproducts of China's breakneck expansion",[42] Canfa Wang, a professor at the China University of Political Science and Law, was chosen by Time Magazine as a "Hero of the Environment". Involved in bill drafting of several important environmental laws and regulations including the amendment of the Water Pollution and Control Law, Professor Wang and his centre have actively participated in environmental lawmaking at the national level.

It is observed that the Chinese public has been speaking louder and making their voice heard by government through environmental NGOs.[43] Nonetheless, Chinese environmental NGOs are faced with more difficult challenges than their western counterparts with institutional obstacles being the main challenge. They are subject to the governance of the 1998 Regulation on Registration and Management of Social Organisations, which poses restrictive registration requirements upon the grass-roots NGOs. The regulation requires NGOs to receive approval from a government organisation before applying for their NGO registration to the civil affairs department.[44] Such approval is practically difficult to obtain as government organisations are rarely willing to take the supervision responsibility. For NGOs who have managed a successful registration, they have to accept the double layered management.[45]

The registration obstacle may have contributed to the predominance of government-organised NGOs among all the Chinese environmental NGOs, accounting for around 37% of the total number.[46] These government-organised NGOs usually have better communication with government and receive more reliable financial support than grass-roots NGOs, which at the same time may compromise their independence and political neutrality.[47] Comparatively, grass-roots NGOs represent a more substantial advance in independence from government, whilst their approving government organisation has nominal oversight for their operational activities and membership.[48] Accordingly, under a high degree of government control, the ability

---

[41] Austin Ramzy, "Wang Canfa, Heroes of the Environment," *Time*, Special Double Issue, October, 2007, http://blogs.guardian.co.uk/environment/2007/10/heroes_of_the_environment.html (accessed September 25, 2008).

[42] Ibid.

[43] All-China Environment Federation, *Summary Report on the Development of China's Environmental Civil Society Organisations in 2008.*

[44] 1998 Regulation on Registration and Management of Social Organisations (China State Council), http://www.lawinfochina.com/display.aspx?lib=law&id=5613&CGid= (accessed May 21, 2017), art. 9.

[45] Percival and Huiyu Zhao, "The Role of Civil Society in Environmental Governance in the United States and China".

[46] All-China Environment Federation, *Summary Report on the Development of China's Environmental Civil Society Organisations in 2008*, 1.

[47] Zhan and Tang, "Political Opportunities, Resources Constraints and Policy Advocacy of Environmental NGOs in China".

[48] Economy, *The River Runs Black*, 305.

of Chinese environmental NGOs in representing the public interests and lobbying against environmental policy remains limited.[49]

The currently limited roles of environmental NGOs in environmental policy may be due to the fact that they are still in the early stage of development and one could optimistically expect Chinese environmental NGOs to increasingly build their capacity and strengthen their influence in monitoring environmental protection, challenging existing legal and policy system, and influencing law and policy reform.[50] Consistent with the political and economic reform of China, civil society would continue to grow and take over more public services functions traditionally undertaken by government.

Internationally, environmental NGOs already play a critical role in water management and allocation. In the US, NGOs may acquire water rights for environmental purposes through market trade and donation although they must transfer the environmental water entitlements to the states as they are prohibited from holding such rights.[51] In Australia, environmental NGOs in New South Wales and South Australia have the legal capacity to hold access entitlements for environmental water allocations.[52] However, in the current institutional framework of China, there is little hope for Chinese NGOs to play a similar role to their western peers in environmental water management without fundamental institutional transformation.

3.3 China's Ongoing Water Policy Reform

3.3.1 Building a Water Saving Society

China's water crisis is far beyond a physical crisis, but a crisis of governance. Dominated by the philosophy of water use "man can conquer nature", China's water policy was traditionally oriented towards construction and utilisation (project oriented) rather than conservation and protection (resource oriented).[53] As intensity in competition between different water values and uses keeps growing, how to allocate limited water resources in a sustainable and efficient manner is of increasing significance. Reflecting on the failure of the traditional principle of water use, and particu-

[49] Zhan and Tang, "Political Opportunities, Resources Constraints and Policy Advocacy of Environmental NGOs in China".

[50] Percival and Huiyu Zhao, "The Role of Civil Society in Environmental Governance in the United States and China".

[51] Alex Gardner and Kathleen H. Bowmer, "Environmental Water Allocations and Their Governance," in *Managing Water for Australia: The Social and Institutional Challenges*, ed. Karen Hussey and Stephen Dovers (Collingwood: CSIRO Publishing, 2007), 43–58.

[52] Ibid.

[53] Shucheng Wang, *Resource-oriented Water Management: Towards Harmonious Coexistence between Man and Nature,* 2nd ed. (Singapore: World Scientific Publishing; Beijing: WaterPower Press, 2006), 37.

larly its perception on the human-nature relationship, the Chinese government has initiated an ambitious water reform "Building a water-saving society", which aims to achieve "harmonious coexistence between man and nature" by establishing a national water right system.[54]

The concept of "Building a water-saving society" was first introduced by the central committee of the Chinese Communist Party in its advice to the national "10th Five Year Plan (FYP) (2000–2005)" in 2000. In the context of China's traditional water institutional arrangements, water resources were administratively allocated. As mentioned earlier, the State Council's 1993 Implementing Regulation on Water Abstraction Permits explicitly prohibited water abstraction permits transfer (Article 26). Nevertheless, as the gap between the increasing water demand and limited water supply kept widening, the administrative instrument for water allocation faced its biggest challenge and failed to allocate water to its higher valued uses and effectively solve water conflicts. As a result, the Chinese government began to explore the market-based mechanisms in water allocation. In October 2000, the then Minister of Water Resources, Mr. Wang Shucheng, presented his speech entitled "Water Rights and Water Markets: Economic Measures for Achieving Optimal Allocation of Water Resources" at the annual conference of the Chinese Hydraulic Engineering Society, and called upon the application of market-based water allocation mechanisms, especially water rights trading.[55] One month after the minister's speech, two cities in Zhejiang Province, Dongyang and Yiwu, came to an agreement, through which Yiwu bought a permanent water use right of 50 million cubic metres of Dongyang's Hengmian Reservoir by paying Dongyang RMB 200 million yuan.[56] The validity and legitimacy of this trade remained controversial as there was no legal foundation for water rights trading at the time, but the MWR expressed a supportive attitude towards the trade intending to explore the experiment of water rights trading in China. Such government-driven water trade broke new grounds for China's administrative water allocation tradition and was viewed as the very first water trading in China.[57]

The 2002 amended Water Law translated the "Building a water saving society" policy into legislation in its Article 8:

> The state encourages strict economy on the use of water, greatly promotes water-conserving measures, spreads the use of new technologies and techniques for water-conserving, develops water-conserving industries, agriculture and services, and builds a water-saving society…[58]

However, the 2002 Water Law did not take the opportunity to go further on water rights and water trading as the reform was still in the early exploratory stage.

[54] Ibid.

[55] Ibid.

[56] Manhong Shen, *The Studies of the Institutions of Water Rights Transactions: The Case Studies in China* (Hangzhou: Zhejiang University Press, 2006), 57.

[57] Ibid.

[58] China 2002 Water Law, art. 8.

3.3.2 Establishing a Water Rights System

Since 2002, the MWR and local governments launched a number of pilot programmes to implement the water-saving policy. The MWR's 11th FYP for Water Conservancy highlighted it as one of the major priorities for the period of 2006–2010. At essence, it involves a transformation of water management regime from a project-based administrative approach to a resources-based, market-oriented approach.[59] The theory of water rights and water markets is viewed as a fundamental policy regime in the reform, which established specific strategies to define initial water rights through water planning, to specify water entitlements through the total quantity control and quota management of water use, and to provide water trading rules to facilitate the development of water trading markets.[60]

Up to 2005, 10 cities and one provincial region were selected as national level pilots for the "Building a water saving society" reform. Pilot programmes were launched and implemented to experiment water rights initialisation and trading, and new regulations and policies were developed to define water property rights. Ningxia Hui Autonomous Region is the first pilot province for the "Building a water saving society" initiative. As an arid area in the Yellow River Basin in the northwest of China, Ningxia faces severe water stress yet with very low water use efficiency. The river basin authority YRCC commenced pilot programmes in water rights transfer from agriculture to industry in Ningxia and Inner Mongolia in 2003. The MWR issued its guidelines on the pilot programmes in 2004 to promote the further development of water rights trading, based on which the YRCC formulated the Trial Implementing Regulation on Water Rights Transfer in the Yellow River Basin.[61] The regulation provided principles for water rights transfer, approval procedures, water transfer planning, duration and expenses of water transfer, etc.[62]

The pilot programmes have demonstrated that water rights trading can be an effective instrument to reallocate limited water resources towards higher valued uses, and a national water rights system needs to be established to govern water rights initialisation and transfer. In 2005, the MWR promulgated "the Framework of Establishing a National Water Rights System", which defines the concept of a water rights system as:

> A water rights system refers to the rules that define, allocate, adjust, protect, and enforce water rights, and clarify the rights, obligations and interests between governments, between governments and water users, and between water users. It is a suite of laws, institutions, and mechanisms, to regulate and protect water rights.[63]

---

[59] Wang, *Resource-oriented Water Management*, 9–10, and 121–122.

[60] Ibid.

[61] 2004 Trial Implementing Measure on Administration of Water Rights Transfer in the Yellow River Basin (Yellow River Conservancy Commission), http://www.yrcc.gov.cn/ziliao/zcfg/guizhang/200612/t20061222_9390.htm (accessed September 26, 2008).

[62] Ibid.

[63] The Framework of Establishing a Water Rights System (China Ministry of Water resources), 2005, http://www.mwr.gov.cn/tzgg/qt/20050117000000645755.aspx (accessed September 26, 2008), Section 2, 3–4.

According to the framework, the water rights system consists of three key institutions: water resources ownership, water resources use rights, and water rights transfer.[64] Such systematic structure is based on the principle that China's water rights system should be a combination of administrative and market-based water allocation mechanisms in the context of China's socialist market economy.[65] It further outlines the key components of each institution as shown in the table (Table 3.3) below.

3.3.3 *Implementing the Strictest Water Resources Management System*

Despite substantial efforts in the 10th and 11th FYP periods, water use security as well as water quality control still remain immense challenges for China. It was recognised that "solution to the increasingly complex water resources problems and realisation of efficient use and effective protection of water resources require good systems, good policies and relevant reform".[66] The State Council further pushed the reform towards more strict water governance. In early 2012, the MWR developed its 12th FYP for Building a Water Saving Society, in which one of the main tasks regarding the water-saving society building in the 12th FYP period (2011–2015) is to implement the most strict water resources management system, and refine its mechanisms of total quantity control of water consumption and quota management.[67]Around the same time, the State Council released its Opinions on the implementation of the strictest water resources management system, through which the "three red lines" policy was announced.[68] The "three red lines" policy intends to strictly control excessive growth of water use and water pollution, and establish clearly defined and binding limits on water use quantity, efficiency, and water quality, with its 2030 goals to: (1) limit the total annual national water consumption below 700 billion cubic metres; (2) reduce water consumption per RMB10,000 yuan industrial value added to less than 40 cubic metres, and increase irrigation use

[64] Ibid.

[65] Ibid.

[66] Briefings on the State Council's Opinions on Implementing the Strictest Water Resources Management System (China Ministry of Water Resources), 2012, http://www.china.org.cn/china/2012-02/17/content_24664350.htm (accessed 23 August, 2014).

[67] The 12th Five Year Plan for Building a Water Saving Society (China Ministry of Water Resources), 2012, Beijing: Ministry of Water Resources.

[68] 2012 China State Council Opinions on the Implementation of the Strictest Water Resources Management System, State Council Document No.3, 2012, http://www.gov.cn/zwgk/2012-02/16/content_2067664.htm (accessed October 24, 2016).

Table 3.3 China's water rights system framework

Water resources ownership	Water use rights	Water rights transfer
The state ownership and unified water management including total quantity control and quota management, and national water planning;	Allocation of water use rights including river basin water allocation, total quantity control of water consumption, quota management, registration and administration of water rights, consultation, and monitoring;	Administration of water rights transfer, regulation of water transfer contracts, proclamation, negotiation and consultation, and protection of third parties;
River basin coordination and regional dispute arbitration mechanisms;	Administration of water abstraction, including enforcement of water abstraction permits, and the paid water resources use system;	Establishment of water markets.
Evaluation of water resources including assessments of economic, environmental, and ecological values of water resources, estimation of water availability, and trend analysis;	Conservation and protection of water and water environment;	
Governance of inter-basin water transfer;	Legal protection of water use rights.	
Decision making and supervision of water allocation.		

efficiency to above 0.6; and (3) reach 95% of tested water meeting national water quality guidelines including requirements on a range of major pollutants.[69]

The State Council's Opinions highlighted six measures in order to achieve these goals, which are highly in line with the national water rights system framework developed in 2005[70]:

- Strictly implement water planning and enhance water resources assessment;
- Strictly control total quantity of water consumption in river basins and administrative regions;
- Strictly implement the water abstraction permit system;
- Strictly practice paid use of water resources;
- Enhance management and protection of groundwater;
- Reinforce unified management of water resources.

It implies that where the total water consumption and abstraction has reached or exceeded the defined total quantity limit, no further water abstraction permits should be issued or approved. Water trading then becomes an alternative vehicle to obtain

[69] Ibid.

[70] Ibid.

water use rights. The State Council's Opinions therefore called for market-based water allocation mechanisms, to further develop and refine the water rights system, establish water markets, encourage and facilitate water trading.[71]

3.3.4 Nationwide Water Rights Piloting

In July 2014, the MWR commenced its nationwide water rights piloting programme, which included seven provinces/autonomous regions across the country. Based on the previous piloting experiences, these seven provincial areas were selected to explore the verification and registration of water rights, water rights transfer models, and corresponding legal and policy developments. The following table (Table 3.4) shows the details about these seven pilots against their focus areas.[72]

The 3-year piloting programme is expected to complete by the end of 2017. Based on the experiences learnt through the programme, the MWR will develop regulations or guidelines on water rights verification, registration, and trading. To provide guidance in water rights trading practices, the MWR released Provisional Measures on Administration of Water Rights Trading in April 2016. Its Article 2 defines water rights trading as:

> Where the water use rights are reasonably defined and allocated, behaviours that achieve water use rights transfer between regions, basins, upstream and downstream, industries, and individual users through market mechanisms.[73]

Three types of water rights trading are further listed in Article 3[74]:

- Regional water rights trading: between local governments at or above the county level or departments or units authorised by them, on surplus water within the water use total control quota and allocation quota, within a same basin, or between different basins with appropriate conditions for water transfer.
- Water abstraction rights trading: Entities or individuals who have obtained water abstraction rights (excluding urban public water supply enterprises) conserve water resources through adjustments in product and industrial structures, technology innovation or other water-saving measures, and transfer with compensation, the part of their water abstraction rights on the conserved water, within the duration and quota of their water abstraction permits.
- Water rights trading between irrigators: that may be water users' organisations or individual users of irrigation water with clearly defined water rights.

---

[71] Ibid.

[72] Workplan for Water Rights Piloting (China Ministry of Water Resources), 2014, Beijing: Ministry of Water Resources.

[73] 2016 Provisional Measures on Administration of Water Rights Trading (China Ministry of Water Resources), http://www.gov.cn/zhengce/2016-05/22/content_5075679.htm (accessed May 27, 2017), art. 2.

[74] Ibid, art. 3.

Table 3.4 Nationwide water rights piloting against focus areas

Focus	Pilots	Key tasks
Water rights verification and registration	Ningxia Hui Autonomous Region	Verify agricultural water use rights in the irrigation districts served by water diversion from the Yellow River
		Explore water registration options
		Further complete water rights verification for other water sources including both surface water and groundwater
		Establish database for water rights registration
		Develop legal and policy documents for water rights verification and registration
	Jiangxi Province	Register and verify water abstraction rights of water abstraction permit holders
		Explore various registration options
		Register and verify water use rights for water resources from ponds or reservoirs of rural collective economic organisations
	Hubei Province	Conduct water resources surveys of ponds and reservoirs of rural collective economic organisations
		Register and verify water use rights for water resources from ponds or reservoirs of rural collective economic organisations which have completed the property rights reform
		Establish the water rights verification and registration system
Water rights transfer models	Inner Mongolia Autonomous Region	Cross-city water rights trading
		Establish the water rights trading platform
		Explore the operation mechanisms for water rights trading including trading rules, pricing, and third party impact assessments
	Henan Province	Cross-basin water rights trading
		Establish the water rights trading platform
		Develop information systems, trading rules, and risk control mechanisms
	Gansu Province	Initial allocation of water rights, and water rights verification and registration
		Establish the water rights trading platform
		Explore water rights trading between different water users and industries
		Develop water rights trading system
	Guangdong Province	Establish the water rights trading platform
		Explore water rights trading between upstream and downstream water users
		Develop information management systems for water rights trading
		Develop regulatory systems for water rights trading

The new Provisional Measures on Administration of Water Rights Trading therefore has established a framework of water rights at three levels, namely the regional level, the water abstraction permit level, and the end-user level. General rules for these three types of water rights trading are also provided in the newly released ministerial regulation. Compared to the 2006 Regulation on Administration of Water Abstraction Permits and Water Resources Fee Collection, the 2016 regulation has broadened the types of tradable water rights, with an intention to foster water rights trading markets across the country. Local governments at provinces, autonomous regions, and municipalities directly under the State Council may formulate specific implementing measures in accordance with these provisions and their local realities.[75] The 2016 Provincial Measures on Administration of Water Rights Trading therefore can be viewed as not only a major reflection on the water rights trading practices in the past 16 years, but also a substantial policy development to provide guidance for the nationwide water rights trading piloting exercise.

3.4 Concluding Remarks

As one of the key priorities under the water reform in China, the establishment of the national water rights system has seen major legal, policy and institutional developments. With the enactment of the new water abstraction regulation, old water abstraction permits are being reviewed, and total water consumption control being gradually specified at both river basin and administrative unit levels. Some areas have been running pilot programmes for water rights initialisation. The new Property Law has explicitly recognised water abstraction rights. The recent Provisional Measures on Administration of Water Rights Trading have established a three-level water rights framework. These developments have provided a primary framework for establishing a national water rights system. Nonetheless, there is still a lack of sufficient legal and policy basis for water rights trading in the existing water governance framework of China. By exploring further implementation of the market-based water allocation mechanisms in China, this book aims to contribute to closing the major governance gaps with the focus areas listed in Table 3.5.

Based on a reflection on the current practices in water rights trading in China, dealt with next in Chap. 4, the chapters thereafter will critically analyse the recent legal and policy developments in China towards tradable water rights including initialisation (Chap. 5), definition (Chap. 6), and administration of tradable water rights (Chap. 7), and administration of water rights trading and establishment of water markets (Chap. 7). The chapters will examine the deficiencies of the current systems of water rights arrangements and trading, explore how China may build on the international trends in water rights trading practices (especially in Australia and the US), and propose legal and policy frameworks for defining and administering tradable water rights in China that underpin sustainable water use in the face of

[75] Ibid, art. 30.

Table 3.5 The focus of this book in the context of China's water rights system framework

Water resources ownership	Water use rights	Water rights transfer
Not a focus in this book	Initialisation of water use rights	Administration of water rights trading
	Definition of tradable water rights	Establishment of water markets
	Administration of tradable water rights	

exacerbated water scarcity, variability, and uncertainty. To this end, the book proposes pragmatic strategies for China's water law and policy reform to move towards tradable water rights.

References

Alford, W. P., & Shen, Y. (1997). Limits of the law in addressing China's environmental dilemma. *Stanford Environment Law Journal, 16*, 125–148.

All-China Environment Federation. (2008). *Summary Report on the Development of China's Environmental Civil Society Organisations in 2008*. Beijing: All-China Environment Federation, 2008.

Australia China Environment Development Program. (2006). *Report on the Development of Water Rights and Trading in the People's Republic of China*. Canberra: Department of Agriculture, Fisheries and Forestry of Australia.

Beyer, S. (2006). Environmental law and policy in the People's Republic of China. *Chinese Journal of International Law, 5*(1), 185–211.

China Ministry of Water Resources (MWR). (2012). *Statistic bulletin on China water activities*. Beijing: China Water and Power Press.

Economy, E. C. (2004). *The river runs black: The environmental challenge to China's future*. Ithaca, N.Y.: Cornell University Press.

Gardner, A., & Bowmer, K. H. (2007). Environmental water allocations and their governance. In K. Hussey & S. Dovers (Ed.), *Managing water for Australia: The social and institutional challenges* (pp. 43–58). Collingwood: CSIRO Publishing.

Jiang, M., Cao, D., & DeLacy, T. (2011). Conflict and coordination of multiple values of water resources: A legal and policy analysis from a perspective of fresh water based tourism in China. In M. Ye, D. Z. Zhu, & T. Zhang (Ed.), *Conference proceeding of the international conference on drinking water safety, security, and sustainability* (pp. 125–138). Hangzhou: Zhejiang University.

Li, T. (2007). Environmental NGOs: Transition from supporting role to leading role. *Journalist Observation*. http://hxtc.china.cn/environment/2007-07/02/content_8467000.htm. Accessed 25 Sept 2008.

Nickum, J. E., & Lee, Y.-S. F. (2013). Same longitude, different latitudes: Institutional change in urban water in China, North and South. In N. T. Carter & A. P. J. Mol (Ed.), *Environmental governance in China* (pp. 83–99). Hoboken: Taylor and Francis.

People's Net. (2006). Water pollution map made by environmental NGOs to disclosing polluters. *People's Net*. http://www.022net.com/2006/9-14/465750243044915.html. Accessed 25 Sept 2008.

Percival, R. V., & Zhao, H. (2014). The role of civil society in environmental governance in the United States and China. *Duke Environmental Law & Policy Forum, 24*, 141–182.
Ramzy, A. (2007, October). Wang Canfa, Heroes of the environment. *Time*, Special Double Issue. http://blogs.guardian.co.uk/environment/2007/10/heroes_of_the_environment.html. Accessed 25 Sept 2008.
Shen, M. (2006). *The studies of the institutions of water rights transactions: The case studies in China*. Hangzhou: Zhejiang University Press.
Sun, F., Yang, Z., & Huang, Z. (2014). Challenges and solutions of urban hydrology in Beijing. *Water Resources Management, 28*, 3377–3389.
Wang, S. (2006). *Resource-oriented water management: Towards harmonious coexistence between man and nature* (2nd ed.). Singapore/Beijing: WaterPower Press/World Scientific Publishing.
World Bank. (2001). *China: Air, land and water: Environmental priorities for a new millennium*. Washington, DC: World Bank.
Wouters, P., Hu, D., Zhang, J., Dan Tarlock, A., & Andrews-Speed, P. (2004). The new development of water law in China. *University of Denver Water Law Review, 7*, 243–308.
Xinhua Net. (2005). The premier Wen Jiabao gave instructions to speed up the assessment of the Nujiang River Project. *Xinhua Net*. http://www.china5e.com/news/water/200509/200509120049.html. Accessed 25 Sept 2008.
Zhan, X., & Tang, S. (2013). Political opportunities, resources constraints and policy advocacy of environmental NGOs in China. *Public Administration, 91*(2), 381–399.
Zuo, L. (2006). Protecting Nujiang River: No Winners Yet. *New Century Weekly*. http://www.china5e.com/news/huanbao/200606/200606200051.html. Accessed 25 Sept 2008.

Legal and Policy Sources

1982 Constitution of the People's Republic of China (2004 Amendments). http://www.npc.gov.cn/englishnpc/Constitution/node_2825.htm. Accessed 21 May 2017.
1984 Law of the People's Republic of China on Prevention and Control of Water Pollution Prevention (2008 Amendments). http://www.mwr.gov.cn/english/laws.html. Accessed 22 May 2017.
1988 Water Law of the People's Republic of China. http://english.mofcom.gov.cn/article/lawsdata/chineselaw/200211/20021100053796.shtml. Accessed 21 May 2017.
1989 Environmental Protection Law of the People's Republic of China. http://www.npc.gov.cn/englishnpc/Law/2007-12/12/content_1383917.htm. Accessed 21 May 2017.
1993 Implementing Measures on the Water Abstraction Permits System (China State Council). http://www.chinawater.net.cn/law/W03.htm. Accessed 28 Sept 2008.
1998 Regulation on Registration and Management of Social Organisations (China State Council). http://www.lawinfochina.com/display.aspx?lib=law&id=5613&CGid=. Accessed 21 May 2017.
2002 Water Law of the People's Republic of China. http://www.npc.gov.cn/englishnpc/Law/2007-12/12/content_1383920.htm. Accessed 21 May 2017.
2003 Administrative Permission Law of the People's Republic of China. http://www.npc.gov.cn/englishnpc/Law/2007-12/05/content_1381966.htm. Accessed 22 May 2017.
2004 Trial Implementing Measures on Administration of Water Rights Transfer in the Yellow River Basin (Yellow River Conservancy Commission). http://www.yrcc.gov.cn/ziliao/zcfg/guizhang/200612/t20061222_9390.htm. Accessed 26 Sept 2008.
2006 Regulation on Administration of Water Abstraction Permits and Water Resources Fee Collection (China State Council). http://www.gov.cn/zwgk/2006-03/06/content_220023.htm. Accessed 21 May 2017.

2007 Property Law of the People's Republic of China. http://www.npc.gov.cn/englishnpc/Law/2009-02/20/content_1471118.htm. Accessed 21 May 2017.

2012 China State Council Opinions on the Implementation of the Strictest Water Resources Management System, State Council Document No.3, 2012. http://www.gov.cn/zwgk/2012-02/16/content_2067664.htm. Accessed 24 Oct 2016.

2016 Provisional Measures on Administration of Water Rights Trading (China Ministry of Water Resources). http://www.gov.cn/zhengce/2016-05/22/content_5075679.htm. Accessed 27 May 2017.

Briefings on the State Council's Opinions on Implementing the Strictest Water Resources Management System (China Ministry of Water Resources) (2012). http://www.china.org.cn/china/2012-02/17/content_24664350.htm. Accessed 23 Aug 2014.

The 12th Five Year Plan for Building a Water Saving Society (China Ministry of Water Resources). (2012). Beijing: Ministry of Water Resources.

The Framework of Establishing a Water Rights System (China Ministry of Water Resources) (2005). http://www.mwr.gov.cn/tzgg/qt/200501170000000645755.aspx. Accessed 26 Sept 2008.

Workplan for Water Rights Piloting (China Ministry of Water Resources). (2014). Beijing: Ministry of Water Resources.

Chapter 4
Alternative Water Governance Mechanisms in China: Examination of Current Practices

Abstract This chapter reflects on the current practices in water rights trading, in the context of other alternative water allocation mechanisms in China. Assisted by empirical analysis, such reflection is important to show that while the merits of water rights trading have been demonstrated through the piloting practices, its effectiveness has also been impeded due to the absence of basic legal and policy infrastructure. More importantly, major governance gaps are identified in the current systems for sufficient facilitation of water markets. Together with the previous chapter (Chap. 3), this chapter serves as a starting point for law and policy reform proposals that will be developed throughout the book in the following chapters.

Keywords Water allocation mechanisms • South-to-North water diversion project • Administrative water allocation • Water pricing • Water rights trading

4.1 Supply-Side Approaches to Water Management and Allocation

A shift in the water management paradigm was internationally observed in the late 1990s to the early 2000s, including "a shift away from sole, or even primary, reliance on finding new sources of supply to address perceived new demands".[1] As discussed in the previous chapters, China is currently transforming its approach to water management from project based to resource based as part of its water reform. Within the traditional water governance framework of China, the logic of solving water shortages is to provide new water supplies to meet growing demands. Such a supply-side approach relies on physical and technical solutions, leading to constructions of new reservoirs and water diversion/transfer projects. As a result, thousands of water engineering projects have boomed dramatically in China since 1949. Among numerous water works, the South-to-North Water Diversion Project (SNWDP) is the most controversial example of supply-side water governance in

[1] Peter H. Gleick, "The Changing Water Paradigm: A Look at Twenty-first Century Water Resources Development," *Water International* 25, no. 1(2000): 127–138, 127.

M. Jiang, *Towards Tradable Water Rights*, Global Issues in Water Policy 18,
DOI 10.1007/978-3-319-67087-4_4

China.[2] This inter-basin water transfer is based on what is believed that China's water resources are unevenly distributed between the south and the north.[3] Northern China faces extreme water shortages to a point where it is unable to sustain its social and economic development, while the south is relatively abundant in water and sometimes has even too much (causing flooding problems).[4] Accordingly, the SNWDP is designed to transfer water from the Yangtze River in the south to the "Three H" area of northern China.[5]

The SNWDP was initially conceptualised as early as the 1950s.[6] Chairman Mao raised this idea when he visited the Yellow River in 1952.[7] From then on, research had been continuously undertaken to assess the project's feasibility and design. Because at that time the government had no financial capacity to afford such giant project, it was not formally put into the central government's agenda until the 1990s. In 1991, the project was highlighted as a policy priority in the 8th Five Year Plan (FYP) period, which ended up being delayed.[8] The 10th FYP in 2001 again highlighted the project after north China had experienced seven consecutive years of drought.[9] Over several decades of research and more than 5 years of intensive planning, the SNWDP Comprehensive Plan was approved by the State Council in 2002.[10] In December 2002, the world's largest inter-basin water transfer project in human history eventually started construction in China.[11]

According to the Comprehensive Plan, the project will cost a total of RMB 500 billion yuan to transfer 44.8 billion cubic metres of water per year to the northwest and the north to relieve their water shortage and ecological degradation problems.[12] In order to achieve this, three water transfer sub-projects, namely, the Eastern Route,

[2] For example, see Jon Barnett et al., "Transfer Project Cannot Meet China's Water Needs," *Nature* 527 (2015): 295–297.

[3] For further elaboration on "the imaginary of a water-rich south versus a water-scarce north" in the context of the South-to-North Water Diversion Project, see Michael Webber et al., "The South-to-North Water Transfer Project: Remaking the Geography of China," *Regional Studies* 51, no. 3 (2017): 374.

[4] Darrin Magee, "Moving the River? China's South-North Water Transfer Project," in *Engineering Earth: The Impacts of Magaengineering Projects*, ed. Stanley D. Brunn (Dordrecht, Heidelberg, London, New York: Springer, 2011), 1500.

[5] For details about the water scarcity in the "Three H" area of Northern China, refer to Sect. 2.1.

[6] Magee, "Moving the River?" 1501.

[7] Bureau of Planning, Design and Management of the South-to-North Water Diversion, "Significant Events of the South-to-North Water Diversion Project," *China Water Resources* 1, Part B (2003): 86–87.

[8] Ibid.

[9] Ibid.

[10] Ibid.

[11] Ibid. Also see Magee, "Moving the River?" 1500.

[12] Bureau of Planning, Design and Management of the South-to-North Water Diversion, "Introduction to the Overall Plan of South to North Water Transfer Project," *China Water Resources* 1, Part B, (2003): 11.

the Middle Route, and the Western Route respectively divert water from the lower, middle, and upper reaches of the Yangtze River.[13]

Through the SNWDP, 44 cities in northern China expect to use fresh water sourced from more than one thousand kilometres away.[14] Four major rivers of China, i.e. Yangtze, Yellow (or Huang), Huai, and Hai will be connected into one integral water network.[15]

The Eastern and Middle Routes are proposed to serve the Three H Plain and the Jiaodong Area, which contribute to the 80% of the Three H Basin's total water shortage.[16] The Eastern Route came online in November 2013 after 11 years of construction, followed by the Middle Route, officially drawing water from Yangtze to the arid north in October 2014.[17] The most controversial Western Route is yet to be commenced due to significantly potential environmental, social, and cultural impacts, technical difficulty, and economic infeasibility.[18]

Although the necessity and urgency of constructing the SNWDP has already been widely recognised and thus becomes the basis of the decision making, the controversy around the project remains prominent at both domestic and international levels.[19] Some observers argue that the SNWDP is another example among China's many other megaprojects, reflecting the nation's "perceived gusto for gigantism".[20] The major question points to the project's detrimental impacts on the environment and ecosystems in both the water importing areas and the exporting areas.[21] Generally speaking, the water receiving areas, such as Beijing, Tianjin, Hebei Province, and the Three H Basin would benefit from the water transfer. The primary objective of the transfer would be to meet industrial, municipal, and agricultural water demands of the north, while ecological demands would also be taken into consideration but without clear specification.[22] More importantly, undesirable aquatic species may be introduced into the northern river systems through the water transfer, which may result in adverse effects on biodiversity of water environments.[23] Water exporting areas apparently face more significant potential in environmental,

---

[13] Ibid.

[14] Ibid.

[15] Ibid.

[16] Ibid.

[17] Office of the South-to-North Water Diversion Project Commission, *Online Media News*, http://www.nsbd.gov.cn (accessed April 12, 2016).

[18] Magee, "Moving the River?" 1506.

[19] Ibid.

[20] Ibid, 1500.

[21] Michael A. Gheleta, "Sustaining the Giant Dragon: Rational Use and Protection of China's Water Resources in the Twenty-First Century," *Colorado Journal of International Environmental Law of Policy* 9 (1998): 241. Also see Changming Liu, "Environmental Issues and the South-to-North Water Transfer Scheme," in *Managing the Chinese Environment*, ed. Richard Louis Edmonds (Oxford, New York: Oxford University Press, 2000), 176.

[22] Magee, "Moving the River?" 1505.

[23] Gheleta, "Sustaining the Giant Dragon".

ecological, social and economic impacts resulting from the construction of the project. The list of potential effects includes sea water intrusion at the mouth and estuary of the Yangtze River,[24] relocation of over 380,000 people,[25] and damage to biodiversity of the river systems.[26]

Another worry about the SNWDP is water pollution. The Eastern Route transfers water through the Grand Canal[27] and watercourses in parallel with it, which connect with the heavily polluted lower reaches of the Yangtze, Huai, Yellow (Huang), and Hai Rivers.[28] To secure water quality above the national surface water quality standards' Level III, billions of yuan was invested in pollution control for the Eastern Route.[29] Hundreds of heavy polluters, especially small chemical paper manufacturers and other small enterprises were shut down to reduce industrial wastewater disposal while wastewater treatment plants were built up to control the pollution. Indeed, the Eastern Route Phase I was far behind schedule as a result of most water quality monitoring cross sections in the system failing to meet the required water quality standard.[30] As the Eastern Route now stands well above the Level III water quality standard, its pollution control remains a long-term challenge. Compared to the heavily polluted waters along the Eastern Route, the Danjiangkou Reservoir, the source of the Middle Route stands better in water quality. However, massive investment is still required to control the water quality of the Middle Route including shutting down heavy polluting businesses, establishing erosion control measures, and constructing additional water treatment plants.[31]

The protection of cultural relics along the routes is another problem. The Eastern Route and the Middle Route go through the "heartland" of China's ancient civilisation.[32] Although the environmental and socio-economic impacts of the SNWDP remain controversial, the project plan was given a go-ahead without an appropriate environmental impact assessment. Cultural heritage conservation only represented a weak voice in the debate.[33] Although the project planning took cultural conservation into account and made revisions to conserve cultural heritage sites, some channels were constructed through precious ancient cultural heritage sites, causing

[24] Liu, "Environmental Issues and the South-to-North Water Transfer Scheme," 180.

[25] Magee, "Moving the River?" 1509.

[26] Gheleta, "Sustaining the Giant Dragon".

[27] The Grand Canal, also known as the Beijing-Hangzhou Grand Canal, is an ancient canal of China, originally constructed as a trade artery.

[28] Magee, "Moving the River?" 1508.

[29] Dengyue Ma, Xuetang Lu, and Lingzhen Zhang, "Current Status and Management of Preexisting Pollution via the Eastern Route of the South-to-North Water Transfer Project," *South-to-North Water Transfers and Water Science & Technology* 5, no. 4 (2007): 30.

[30] Ibid.

[31] Magee, "Moving the River?" 1510.

[32] Chen Shen and Hong Chen, "Cultural Heritage Management in China: Current Practices and Problems," in *Cultural Heritage Management: A Global Perspective,* ed. Phyllis Mauch Messenger and George S. Smith (Gainesville: University Press of Florida, 2010), 77.

[33] Ibid.

unavoidable damage.[34] In fact, cultural conservation-related work did not commence until 2 years after the commencement of the construction of Eastern Route Phase I and Middle Route Phase I.[35] Pressure from politically-driven deadlines combined with lack of funding for cultural conservation led to loss of hundreds of cultural heritage sites and relics in the areas where the Eastern and Middle Routes pass through.[36]

Nevertheless, the biggest challenges to major water projects like the SNWDP, are ones of institutions and governance, rather than those of engineering or environment.[37] The project is highly complex, of which the plan, design, construction, and management work involve a wide range of state-owned enterprises, local governments and agencies at various levels across several provinces, relevant river basin authorities, and central government agencies as well. To respond to the coordination challenge in the SNWDP, the State Council has established a specific commission overseeing major policy matters related to the project. The Commission's Office is equivalent to a ministry level in Chinese bureaucratic hierarchy. In February 2014, the State Council promulgated the Regulation on Administration of Water Supply and Use for the SNWDP. The new regulation is expected to provide a legal framework for management of the project's water supply and use, which emphasises on water quality control and a holistic approach to water allocation planning.[38] Yet, it is a long way ahead for both policy makers and researchers to fully understand the implications of the SNWDP for China's ongoing water governance reform.

4.2 The Administrative Approach to Water Allocation

The administrative approach to water allocation can be understood in a broader sense. For example, the decision making of the SNWDP can be viewed as an administrative water transfer. Instead of such broader perspective, however, this section only refers to, from the demand side of water management, government reallocating water resources from one user to another using administrative instruments.

In many countries around the world, government traditionally controls water allocation.[39] In China, water resources have been completely managed and allocated by government stemmed from its planned economy tradition. China's water

---

[34] Xinhua Net, "Great Impact on Cultural Relics of the South-to-North Project," *Xinhua Net*, 2005, http://www.nsbd.gov.cn/zx/mtgz/20040915/200409150024.htm (accessed September 26, 2008).

[35] Shen and Chen, "Cultural Heritage Management in China," 75.

[36] Ibid, 77.

[37] Magee, "Moving the River?" 1511.

[38] 2014 Regulation on Administration of Water Supply and Use for the South-to-North Diversion Project (China State Council), http://en.pkulaw.cn/display.aspx?cgid=219668&lib=law (accessed May 21, 2017).

[39] Ariel Dinar et al., *Water Allocation Mechanisms: Principles and Examples*, World Bank Policy Research Working Paper, No. 1779, 1997.

legislation explicitly prohibited market water reallocation, except for a few exclusions. China's Constitution and Water Law both state that water resources are a state property. On behalf of the whole people, the government is responsible for managing and allocating water resources within and among regions and sectors in order to achieve the national economic and social goals. When the national development strategy changes, or some specific concerns such as drought or severe environmental problems arise, reallocation of water over space and time, and across different types of uses might be justified. The administrative approach is then to be used to make such modifications. This section provides three examples of China's administrative reallocation of water resources, through which the advantages and disadvantages of the administrative approach to water allocation will be evaluated.

4.2.1 Beijing: Paramount Priority

Beijing is one of the most water-short cities in China with its per capita water availability as low as 193.2 cubic metres, compared to the national per capita water availability of 2093 cubic metres.[40] Beijing's water supply heavily depends on groundwater and natural precipitation as there are no large rivers across Beijing plain.[41] In the last three decades, Beijing has experienced dramatic urbanisation, rapid population and GDP growth, and landscape alteration, which has resulted in an increasingly challenging water shortage.[42] As the capital city of China, Beijing is of unique importance among the country's political, economic, cultural, and educational centres. Accordingly, its water demand is always put in paramount priority by the government.

For several decades, Beijing's two neighbouring provinces, Hebei and Shanxi, have supplied water to Beijing upon the MWR's administrative order, and without market-based compensation, even though such supply has overstretched their own water resources. Hebei Province, for instance, used to share water supply from Guanting Reservior and Miyun Reservior with Beijing since the 1950s.[43] From the early 1980s, Hebei gave up for free its annual water usage quota of 900 million cubic metres from these two reservoirs to Beijing.[44] Between 2003 and 2007, Hebei

---

[40] China National Bureau of Statistics (NBS), *China 2013 Statistical Yearbook*, 2014, http://www.stats.gov.cn/tjsj/ndsj/2013/indexch.htm (accessed October 24, 2016).

[41] Feng Sun et al., "Challenges and Solutions of Urban Hydrology in Beijing," *Water Resources Management* 28 (2014): 3377–3389.

[42] Ibid.

[43] Jie Hu. "Fight for Water Rights of the Juma River between Beijing and Hebei," *New Beijing Newspaper*, 2004, http://www.envir.gov.cn/info/2004/2/225855.htm (accessed September 26, 2008).

[44] Ibid.

undertook drought emergency water supply for Beijing together with Shanxi Province, contributing to a total amount of more than 300 million cubic metres.[45]

Besides direct water supply, Hebei undertook industry restructuring and ecological restoration for water conservation to ensure water quality supplied to Beijing. Such responsibilities added pressure on local economic development and exacerbated poverty of local people in Hebei Province. Chicheng County in Hebei, for example, is one of the poorest counties in China. Yet, all of the county's annual runoff of 347 million cubic metres was supplied to feed Beijing's reservoirs by reducing 90% of its rice cultivation acreage to save local water use.[46] Grazing and livestock husbandry were also completely prohibited to prevent water pollution, resulting in an annual income loss of local farmers as much as RMB 50 million yuan.[47] The county did not acquire its very first compensation of RMB 30,000 yuan until 2004 after more than three decades of sacrifice for Beijing's water supply.[48]

Within the context of the water policy reform, the tradition of sacrifice for the capital city started being questioned with the introduction to the concept of water rights. In 2006, the two governments of Beijing and Hebei signed a memo, in which Hebei was committed to supply of 400 million cubic metres of water for the Beijing 2008 Olympics.[49] In return, Beijing was committed to an investment of RMB 100 million yuan to support Hebei's agricultural production shifts and ecological restoration.[50] Hebei farmers, who shifted from rice cultivation to other less water intensive crops, will be partially compensated for their income loss.[51] In early 2008, the State Council released the water allocation plan for the main stream of Yongding River, a watercourse of the Hai River system, to define the water shares between Beijing and its upstream provinces of Shanxi and Hebei and another neighbouring municipality, Tianjin city. This plan provides with the equally thirsty neighbours of Beijing policy foundation for defined water supply from Yongding River's main stream, although one of the political intentions of the plan was to maintain the river and feed Beijing's reservoirs.[52]

---

[45] Xinhua Net, "Hebei Supplies Water for Beijing in Spite of Its Consecutive Drought for 11 Years," *Xinhua Net*, 2007, http://www.yznews.com.cn/zhxw/2007-09/28/content_1038853.htm (accessed September 26, 2008).

[46] Anonymous, "Exploring the Water Supplier of Beijing, Chicheng County in Hebei: Hardly Compensated for Its Water Supply," *New Beijing Newspaper*, 2005, http://www.jxgdw.com/jxgd/news/gnxw/userobject1ai612675.html (accessed September 26, 2008).

[47] Ibid.

[48] Ibid.

[49] Jun Yang, "Water Supply for Olympics Has Led to Water Conflict," *Nan Feng Chuang*, 10/12/2006, http://www.cnwaternews.com/newscenter/2006/12month/200612102.htm (accessed September 26, 2008).

[50] Ibid.

[51] Xinhua Net, "Hebei Supplies Water for Beijing in Spite of Its Consecutive Drought for 11 Years".

[52] Martin Cosier and Dajun Shen, "Urban Water Management in China," *International Journal of Water Resources Development* 25, no. 2 (2009): 249–268.

4.2.2 *Ecological Emergency*

Another example of administrative water allocation is ecological emergency water transfer. As described earlier in Chap. 2, many water systems of China, especially those in the arid north, are nearly exhausted due to the extremely high utilisation ratios. To rescue the deteriorating ecosystems including drying rivers and shrinking wetlands, the Chinese government has undertaken administrative water transfers to restore ecosystems at the lower reaches of those damaged rivers. Ecological emergency water transfers have been used in northern rivers such as Yellow River and Hei River, to address environmental deterioration in their downstream areas. Through administrative orders of ecological emergency water transfer, water supply to the upstream users is shut down for a period of time, to ensure water flow to the lower reaches.

The restoration of Tarim River is a successful example. Located in Xinjiang Uygur Autonomous Region, Tarim River is the longest inland river of China. Because of high evaporation rates and low precipitation in the basin area, Tarim's ecosystem is highly fragile. From the 1970s onwards, irrigation land in the upstream area was dramatically expanded along with population growth and economic development. Over 300 kilometres of Tarim's downstream, including its end point, Taitema Lake, dried up for 30 years, accounting for one fourth of the river's total length.[53] As a result, all the natural vegetation around Taitema Lake was almost swallowed by neighbouring deserts.[54] From 2000 to 2005, the Chinese government ordered seven ecological emergency water transfers to rescue the degrading ecosystem of Tarim's lower reaches.[55] A total of 2.45 billion cubic metres of water was supplied by upstream users to hydrate the dying ecosystem in the lower reaches.[56] Five of those transfers reached Taitema Lake and restored it after several decades of dry-up.[57] The ecosystem around the lake showed a remarkably positive response to the emergency hydration including vegetation and biodiversity recovering, aquifer level increasing, and groundwater quality improving.[58] Another interesting effect of the ecological emergency water transfers is increased water use efficiency in the upstream areas. The compulsory water transfers have effectively stimulated investments in water saving technologies.

---

[53] Yuelu Zhou, "Study on Ecological Emergency Water Transfer for Tarim Low Stream," *China Water Resources* 2 (2002): 68.

[54] Qingping Zhu et al., "Ecological Water Transfer of China's Inland Rivers," *China Water Resources* B03 (2003): 50.

[55] Zhicheng Jia, "Integrated River Basin Management Makes Taitema Lake Green Again," *China Water Resources Newspaper*, 07/07/2006, http://www.mwr.gov.cn/ztpd/2007ztbd/2006ndslhxw/jszgxxwzpj/20070301105420f8f302.aspx (accessed September 26, 2008).

[56] Ibid.

[57] Ibid.

[58] Zhu et al., "Ecological Water Transfer of China's Inland Rivers".

4.2.3 Less Favoured Agriculture

As a traditional industrial sector, agriculture plays a unique role in China's economic and social development. Along with rapid urbanisation since the 1980s, there are still more than 600 million rural people making a living in the agricultural sector, accounting for over 45% of the total population.[59] More importantly, agriculture is of great significance to China's and even the world's food security. Indeed, China feeds more than 20% of the world's population with only 7% of the world's cultivated land.[60] Food self-sufficiency is the central goal of China's agricultural policy. If China failed to maintain a relatively high level of food self-sufficiency, the potential increases in food import would likely threaten global food security. China is therefore committed to maintaining 95% grain self-sufficiency.[61]

Compared to its contribution to food security, the contribution of agriculture in China to the national GDP is much less significant than those of other industries. The agricultural production of China only accounted for less than 9.2% of the national GDP in 2014 by taking 63.4% of the nation's total water usage.[62] Economic productivity becomes a major drive for the agricultural water use to receive less favour and priority than other uses (municipal and industrial uses). With rapid economic growth and urbanisation, the agricultural share of water use has been steadily decreasing while municipal and industrial uses taking more shares.[63] While biased water allocation policy mostly applies to new sources of water, in some cases industrial and municipal shortfalls are filled by administratively taking water away from agriculture without compensating farmers.

Nevertheless, agriculture remains the biggest water user in China, accounting for 63.1% of the total water use in 2015.[64] Its large amount of water use coupled with low use efficiency means that there is great potential for water saving. The Chinese government have therefore set the "nil growth" goal for agricultural water consumption since 2005. Instead of increasing water supply, the agricultural sector has to deal with its water shortage through water saving technology, production structure shifts, and irrigation infrastructure rehabilitation.

---

[59] NBS, *China 2014 Statistical Yearbook*.

[60] Food and Agriculture Organisation of the United Nations (FAO), *Poverty Alleviation and Food Security in Asia: Lessons and Challenges*, Regional Office for Asia and the Pacific, Bangkok: Food and Agriculture Organisation of the United Nations, 1999, 115.

[61] Bryan Lokmar et al., *China's Agriculture Water Policy Reforms: Increasing Investment, Resolving Conflicts, and Revising Incentives*, Market and Trade Economics Division, Economic Research Service, U.S. Department of Agriculture: Agriculture Information Bulletin Number 782, 2003, 22.

[62] China Ministry of Water Resources (MWR), *2015 Bulletin of China's Water Resources*, 2016, Beijing: Ministry of Water Resources, 4.

[63] Lokmar et al., *China's Agriculture Water Policy Reforms: Increasing Investment, Resolving Conflicts, and Revising Incentives*, 2.

[64] MWR, *2015 Bulletin of China's Water Resources*.

There are legitimate reasons to reallocate water from low valued uses in agriculture to higher valued uses in other industries. Yet, the challenge is whether the instrument used to achieve economic efficiency is effectively taking equity into consideration. The administrative approach without appropriate compensation to agricultural water users would undoubtedly give them little incentive to invest in water saving and increase their water use efficiency. Moreover, it is not fair to take water away from farmers for free, who are more disadvantaged than other communities because their incomes are much lower than that of the urban population.[65]

The three examples analysed above show that the administrative approach to water allocation has its advantages and disadvantages. It is effective in achieving government's political goals at the macro level, such as its success in securing the water demand for the 2008 Beijing Olympic Games. It also works well in fulfilling macro-economic development policies, such as transferring agricultural water to industrial and urban use. On the other hand, the administrative approach to water allocation fails to address the equity issue properly. Both the Beijing case and the agriculture case have shown its failure in equity. As for the case of ecological emergency, the instrument has accomplished equity to some extent between the upstream users and the downstream users, but this was just an additional effect besides its main concern about ecosystem deterioration.

4.3 Other Economic Instruments: Water Pricing, Water Resources Fees, and Sewage Treatment Fees

From the traditional perspective of water allocation and management, water has been viewed as a social or public good and thus has been provided for free or with heavy government subsidies. With increasing water demand and worsening water shortages, however, water availability becomes scarcer, and acquisition of new supplies becomes increasingly difficult and expensive. Free or cheap water supply is therefore unable to encourage water conservation and efficient water use. Government needs to find alternatives to managing the water demand and supply balance.

Water pricing is one of the economic instruments used by government to restrain users' demand and improve water allocation. China's water crisis has revealed that an irrational water pricing policy presents as a major contributing factor to low efficiency of water use and waste in agriculture, households and industry. To respond to the problem, the Chinese government initiated the water pricing reform, which is also an essential part of China's ongoing water policy reform. This section investigates the latest development of China's water pricing policy and current practices of its water pricing reform. Two specific issues, water resources fees and discharge treatment fees, are also addressed in this section because they are included in the

[65] According to China 2014 Statistics Yearbook, there are still over 70 million rural people with an annual income less than RMB 2300 yuan, compared to the national average of RMB 20,167 yuan.

water price in China. The effectiveness of water pricing in water allocation within China's context is to be concisely evaluated by examining the agricultural water pricing reform and the urban water pricing reform.

4.3.1 *From Free Water Use to Administrative Fee Collection to Government-Set or Government-Guided Water Prices*

China's water pricing policy has experienced a shift from free supply to paid water use, and very slowly, its reform goes towards full cost recovery for water supplies. China's water users had not been asked to pay for water services during the first 16 years of new China from 1949.[66] Although the State Council approved the Regulation on Hydraulic Engineering Water Fee Collection, Use and Management formulated by the Ministry of Water Conservancy and Electric Power in 1965, China's water projects hardly charged their users during the period of Cultural Revolution.[67]

The concept of water as an economic good arose with the development of China's economy reform in the early 1980s and government water pricing reform started from this period. In 1985, the Regulation on Hydraulic Engineering Water Fee Calculation, Collection and Management was promulgated by the State Council, which confirmed that water supplies from hydraulic engineering projects should be a paid service and thus an administrative fee should be charged.[68] Most provincial governments issued their implementing provisions in accordance with this regulation. The paid use of water supply was then reconfirmed in Article 34 of the 1988 Water Law, which constituted the legal basis of China's water pricing system. Water supply was considered a paid administrative service from government and was thereby charged as fees rather than prices. The concept of water prices did not appear in China's legislation until 1991 when the MWR formulated the Principles for Calculating Prices of Rural Water Supplies (Trial). Water supplies started being viewed as a commodity in China since then.

According to the 1997 Price Law (Articles 19 and 20), the price of water shall be set or guided by government rather than regulated by market due to water's significance to the nation's economic development and people's life.[69] Local governments

---

[66] Xiaohua Tao, "The Historical Development of China's Water Pricing System," *Zhi Huai* 12 (2004):16–17.

[67] Ibid.

[68] 1985 Regulation on Hydraulic Engineering Water Fee Calculation, Collection and Management (China State Council), http://www.hwcc.com.cn/newsdisplay/newdisplay.asp?id=116993 (accessed September 27, 2008), art. 1.

[69] 1997 Price Law of the People's Republic of China, http://www.npc.gov.cn/englishnpc/Law/2007-12/11/content_1383577.htm (accessed May 22, 2017), art. 19, art. 20.

at or above the county level can determine and adjust water prices through their price control departments and water administration departments.[70]

The 2002 Water Law has established the system of paid use of water resources (Article 7). Its Article 49 has further stipulated that: "Water use shall be metered … water use shall be charged volumetrically and progressive payments for the part in excess of the quota shall be adopted".[71]

In accordance with the 2002 Water Law and the 1997 Price Law, the 2003 Regulation on Administration of Water Supply Prices for Water Conservancy Projects replaced the 1985 Regulation. The following issues in the water pricing system are readdressed in the 2003 regulation[72]:

- Water supply prices of water conservancy projects consist of production costs of water supply, production expenses of water supply, profits, and taxes (Article 4).
- Water supply prices of water conservancy projects shall be managed in a decentralised way under a unified policy. They can be government-set or government-guided depending on different situations (Article 5).
- Water supply of water conservancy projects for different uses shall be charged differently according to the national economic policy and the affordability of users (Article 10).
- Two-part prices, i.e. with fixed and volumetric components, shall be gradually applied for water supply of water conservancy projects. Water use shall be managed on a quota basis, and progressive payments shall be applied for the part in excess of the quota (Articles 13–14).

All these legal developments show the trend in China's water pricing reform that water charges have shifted from heavily subsidised administrative fees into government-set or government-guided prices. Within this context, it is not surprising to see that water prices are climbing in various water using sectors of China.

4.3.2 Climbing Water Prices

Although agriculture, households, and industries all need to pay much more than before for their water use, the water pricing reform has different implications for different water using sectors. By looking at the agricultural water pricing reform as well as the urban water pricing reform, this subsection examines the effectiveness of water pricing on optimising water allocation.

---

[70] Ibid.

[71] 2002 Water Law of the People's Republic of China, http://www.npc.gov.cn/englishnpc/Law/2007-12/12/content_1383920.htm (accessed May 21 2017), art. 49.

[72] 2003 Regulation on Administration of Water Supply Prices for Water Conservancy Projects (China State Council), http://news.xinhuanet.com/zhengfu/2003-07/22/content_987613.htm (accessed September 27, 2008).

As the biggest water user of China, agriculture uses water with low efficiency and this leads to excessive waste. The water saving goal of nil growth for agricultural water use has not only implied the government's priority in water allocation to sectors other than agriculture, but also revealed the great potential for agricultural water use to improve use efficiency. Water pricing has been applied as an economic incentive to encourage farmers to conserve water.

China's agriculture is dominated by individual family-based operations, which makes water metering difficult. Coupled with poor metering facilities, "per unit area" was used as the main water charge method in agriculture water use.[73] In other words, agricultural water was charged per irrigated area. Since the cost of water was not related to the actual volumes used, farmers tended to use basic irrigation patterns such as flood irrigation, which consumes and wastes more water. The most remarkable change in agricultural water pricing is therefore changing the charge method. Volumetric charge of water has been applied to gradually replace per area pricing.[74] In some regions, water charge based on use volume has motivated farmers to reduce the costs for water use by adopting water saving irrigation technologies and investing in maintenance of water delivery channels.[75] However, water is still charged at a flat rate based on land areas in many irrigation districts where volumetric facilities are not available.[76] As a result, the recent State Council General Office's Comments on Promoting the Comprehensive Agricultural Water Pricing Reform suggested to speed up the development of measurement infrastructure in irrigation districts.[77] According to the Comments, by investing in irrigation and water conservation facilities, the Chinese government will move towards a tiered pricing mechanism for agricultural water use.[78]

The economic value of water in agriculture is much lower than for industrial or domestic use. To ensure food security and social stability, however, agricultural use of water is of great significance in China. The pricing principle for agriculture water is that, instead of including profits and tax payment, the agricultural water price only recovers the production costs and expenses for water supplies.[79] The system of water charge collection was also adjusted to avoid the risk of free riding and release

[73] Yongsong Liao et al., *China's Water Pricing Reforms for Irrigation: Effectiveness and Impact*, Comprehensive Assessment of Water Management in Agriculture Discussion Paper 6, Colombo, Sri Lanka: International Water Management Institute, 2008, 13.

[74] Ibid.

[75] Zhi Mao and Yuanhua Li, "Effect of Reform of Water Charge on Water Saving in Irrigation Scheme in China," the 2nd Inter-Regional Conference on Environment-Water, September 1–3, 1999.

[76] Liao et al., *China's Water Pricing Reforms for Irrigation*.

[77] State Council General Office's Comments on Promoting the Comprehensive Agricultural Water Pricing Reform (China State Council General Office), 2016, Beijing: China State Council General Office.

[78] Ibid.

[79] 2003 Regulation on Administration of Water Supply Prices for Water Conservancy Projects, art. 10.

the extra burden on farmers caused by multiple administrative levels.[80] To recover production costs and water supply expenses, the water price per cubic metre will need to be increased from RMB 0.03 yuan to 0.035 yuan.[81] However, the average actual water payment of farmers was 0.05 yuan per cubic metre, 40% higher than the average water price for agricultural use.[82] Although some farmers did not pay that much for their water use thanks to relatively good collection procedures in their local area, others were paying even more than the average amount. In fact, illegal collection and the arbitrary increase in water fees widely exist in rural China as various fees are often collected in the name of water fees at local township or village levels.[83] Compared to the administrative water fee collection in the past, it is now water suppliers who directly collect water charges.[84] As a result, even though the agricultural water price is gradually getting higher and some farmers do have to pay more than before, this change has actually reduced many farmers' water costs. Since economic incentives for farmers to use less water are still lacking, the impact of water pricing reform on farmers' water use behaviours remains limited.

Urban water users of China have been experiencing a speedier rise in water prices than farmers. The urban water pricing reform aims for a full urban water price that not only recovers the costs and expenses of water supply but also includes tax payment and reasonable profits.[85] Some researchers appraise the effectiveness of increased water rates in reducing water demand.[86] On the contrary, others tend to be critical about the urban water price hike, putting the burden on disadvantaged households while high-income families will continue to use as much water they like to maintain their lifestyle regardless of the rising water price.[87] Although a water price rise has to go through public hearing, its legitimacy and fairness are questioned due to its lack of transparency and "fake" representation.[88]

In 2014, a tiered water pricing system for tap water was put in place, under which Beijing residents may be charged as much as RMB 9.00 yuan per cubic metre for their tap water use (Table 4.1). Beijing is not the only city in China that adopts the

---

[80] 2002 National Planning Commission's Advices on Agricultural Water Pricing Reform (China National Planning Commission), http://www.hwcc.com.cn/newsdisplay/newsdisplay.asp?Id=38474 (accessed September 27, 2008).

[81] Ibid.

[82] Ibid.

[83] Liao et al., "China's Water Pricing Reforms for Irrigation".

[84] 2002 National Planning Commission's Advices on Agricultural Water Pricing Reform.

[85] 2003 Regulation on the Administration of Urban Water Supply Prices (China State Council), http://www.china5e.com/laws/index2.htm?id=200405090021 (accessed September 27, 2008), art. 7.

[86] Wenlai Jiang and Qu Tang, "A Study on Water Pricing Reform in Beijing," *Journal of Economics of Water Resources* 27, no. 3 (2009): 30–32.

[87] B. Zhao et al., "To Pay More or Use Less: China's Urban Water Pricing Regime and Water Awareness," in *MODSIM2011: 19th International Congress on Modelling and Simulation*, ed. F. Chan et al. (Perth, WA: The Modelling and Simulation Society of Australia and New Zealand, 2011).

[88] Ibid.

Table 4.1 Tiered water pricing for tap water in Beijing

No. of tiers	Annual water usage per household (cubic metres)	Water price (Yuan per cubic metre)	Components in water price: Tap water rates (Yuan per cubic meter)	Water resources fees (Yuan per cubic metre)	Sewage treatment fees (Yuan per cubic metre)
1	0–180	5.00	2.07		
2	181–260	7.00	4.07	1.57	1.36
3	>260	9.00	6.07		

Modified from Beijing Municipal Commission of Development and Reform, 2014 Implementing Rules of Residential Water Tiered Pricing for Beijing

tiered water pricing system. In late 2013, the Guidelines jointly released by the NDRC and the MOHURD stated the objective to implement the tiered water pricing system and block rate structure for urban residential water use across all the cities in China.[89] How such new development in urban water pricing contributes to water demand management in China remains an open question.

4.3.3 *Water Resources Fees and Sewage Treatment Fees*

In accordance with the principle of paid use of water resources, water resources fees are applied in China. Directly diverting surface water from rivers or lakes and extracting groundwater are payable with water resources fees.[90] In consultation with the Ministry of Finance and the MWR, the NDRC is responsible for setting water resources fee collection standards for water conservancy projects directly under the central government whose water abstraction permits have to be approved by river basin authorities, and water conservancy projects across provincial territories.[91] Taking local water availability and economic growth conditions into consideration, local governments at the provincial level set their local water resources fee standards. In the agricultural sector, the water resources fee standard can be lower than those for other types of use.[92] Agricultural water resources fees are often waived in

[89] Guidelines on Accelerating the Establishment and Improvement of the System of Tiered Water Pricing for Urban Residential Use (China National Development and Reform Commission and the Ministry of Housing and Urban-Rural Development), 2013, http://www.gov.cn/zhengce/2016-05/22/content_5075654.htm (accessed May 22, 2017).

[90] 2006 Regulation on Administration of Water Abstraction Permits and Water Resources Fee Collection (China State Council), http://www.gov.cn/zwgk/2006-03/06/content_220023.htm (accessed May 21, 2017), art. 2.

[91] 2008 Measures on Management of Water Resources Fee Collection and Use (China Ministry of Finance, National Development and Reform Commission, Ministry of Water Resources), http://www.mof.gov.cn/zhengwuxinxi/caizhengwengao/caizhengbuwengao2008/wengao200811qi/200903/t20090304_118518.html (accessed May 21, 2017), art. 8.

[92] 2006 Regulation on the Administration of Water Abstraction Permits and Water Resources Fee

most provinces while a few provinces do collect water resources fees at a much lower rate than for industrial and domestic use.[93] Provincial governments set a quota for direct surface diversion and groundwater extraction for agricultural water use, and farmers pay water resources fees only when such quota is exceeded.[94] Farmers who use water supplied by water conservancy projects pay water prices including water resources fees as a part of the water supply cost.[95] In 2013, the NDRC, the MOF, and the MWR jointly released a guide on minimum water resources fee collection standards that local governments were suggested to meet by the end of the 12th FYP period (the end of 2015).[96] Rather than reflecting water scarcity, water resources fees are more a symbol of the state's ownership over water resources in China. It remains a challenge for the Chinese government to raise water resources fee collection standards to the level that will have an effective impact on changing water use behaviours.

Charged on the basis of users' water consumption, the sewage treatment fee is another component of the urban water price.[97] The MOF, the NDRC, and the MOHURD issued the Measures on Administration of Sewage Treatment Fee Collection and Use in late 2014, which took effect in March 2015. Sewage treatment fees are paid by discharge units and individuals according to the "polluter pays" principle for the construction and operation of urban sewage treatment facilities and the treatment of sludge.[98] Under the Measures, sewage treatment fees shall be levied in cities, counties, and organic towns where sewage treatment plants are in place.[99] Local governments above the county level set the sewage treatment fee collection standards within their jurisdiction according to the principle of covering normal operation of sewage treatment facilities, sludge treatment and disposal costs, and reasonable profit.[100] Although the coverage of sewage treatment facilities has experienced a rapid growth in China in the last 10 years, sewage treatment fee standards remain low, failing to cover the facilities operation costs. A recent research showed that current sewage treatment fees in China only partially cover the facilities operation costs as sewage treatment plants heavily rely on government financial subsidies.[101]

Collection, art. 28.

[93] Haifeng Wang et al., "The Issues in the History of Administration of Water Resources Fee Collection," *Price Monthly* 411 (2011): 42–46.

[94] 2006 Regulation on the Administration of Water Abstraction Permits and Water Resources Fee Collection, art. 33.

[95] Ibid.

[96] 2013 Minimum Standards for Water Resources Fee Collection by the End of the 12th Five Year Plan (China National Development and Reform Commission, the Ministry of Finance, and the Ministry of Water Resources), http://www.ndrc.gov.cn/fzgggz/jggl/zcfg/201301/W020130114400736805045.pdf (accessed May 21, 2017).

[97] 2003 Regulation on the Administration of Urban Water Supply Prices, art. 3.

[98] Ibid.

[99] Ibid, art. 7.

[100] Ibid, art. 12.

[101] Xue Tan et al., "Institutional Analysis of Sewage Treatment Charge Based on Operating Cost of Sewage Treatment Plant – An Empirical Research of 227 Samples in China," *China Environmental Science* 35, no.12 (2015): 3833–3840.

It is widely recognised in literature that as an economic instrument, water pricing could be an effective means to improve water allocation. Unlike those supply-side management instruments such as reservoir construction and water transfers, water pricing theoretically affects water use by changing water users' behaviour. Well-designed water prices send users the right signal and provide them with an incentive to conserve water. Conversely, irrational water pricing misleads users with wrong signal about water scarcity and its true value. In this aspect, China's water pricing reform is of significance in optimising water allocation. The characteristics of water as an economic good have been primarily accepted through the reform. Moreover, although currently water is still well underpriced in China, increasing charges have to some extent played an effective role in use efficiency improvement and water conservation. By charging different prices, government's water allocation policy is implemented in various water using sectors. However, efficient allocation of water is not the only purpose of water pricing. It can also be designed to accomplish other objectives such as fairness and equity, and social stability, which makes the water pricing reform a very sensitive issue. The current practices of China's water pricing reform have clearly indicated the difficulty in increasing water prices to the level that costs of water supplies and services can be fully recovered. Its effectiveness in water allocation optimisation is therefore largely undermined as the water price is unable to reflect its true value.

4.4 Water Rights Trading

This section reviews three representative water trading cases in China: Dongyang-Yiwu water trading, Zhangye "water notes" transfer and water rights transfer in the Yellow River Basin. Having taken place in the period of China's ongoing economic and water policy reforms and strongly supported and guided by government, all these cases are ground breaking.

4.4.1 Water Trading Between Dongyang and Yiwu

In November 2000, two counties of Jinhua City in Zhejiang Province, namely Dongyang and Yiwu, came to an agreement, in which Yiwu purchased the permanent water use right of about 50 million cubic metres per year from Dongyang's Hengjin Reservoir by paying Dongyang RMB 200 million yuan.[102] This deal

[102] Manhong Shen, *the Studies of the Institutions of Water Rights Transactions—the Case Studies in China* (Hangzhou, Zhejiang University Press, 2006), 57.

between the two local governments broke the tradition of administrative water allocation and thus was viewed as the first water trading case in China.[103]

Well known for small commodity manufacture and wholesale, Yiwu is a well-developed economy with rapid population and economic growth which led to increasing demands for water. However, its local water supply was unable to match up such an increase in water demand, as its per capita water availability of 1130 cubic metres is less than half of the national average.[104] As the biggest obstacle for Yiwu's economic and social development, water shortage became the most urgent issue for the government to cope with. Purchasing water rights from its neighbouring county Dongyang appeared to be one solution.

Located in the upstream area of the Jinhuajiang River Basin, Dongyang has more abundant water resources than Yiwu, with its 2126 cubic metres of per capita water availability.[105] With no need to concern about water supply, water use efficiency was fairly low in Dongyang. Furthermore, Dongyang actually had to discharge more than 30 million cubic metres of water in flood season every year to regulate its water supply.[106] The supply capacity of Hengjin Reservoir, which is the biggest in Dongyang, can be further increased. By restoring the reservoir and the connected irrigation infrastructure, Hengjin Reservoir could achieve an increase in supply capacity of 53 million cubic metres.[107] On the other hand, the Zixi River in Dongyang was exploited to bring more water to the reservoir. Through a two-cascade hydropower development, 50 million cubic metres of water can be delivered to Hengjin Reservoir.[108] After making these two investments, Dongyang had more than 100 million cubic metres of extra water availability.

As Yiwu's upstream neighbour within the same river basin, Dongyang used to voluntarily help Yiwu in water shortage. For example, it provided more than two million cubic metres of water in 1995 and 1996 respectively through administrative negotiations coordinated by their superior government.[109] However, the administrative approach to water allocation between the two counties was not reliable for Yiwu as the negotiation process was time-consuming and economically costly, and most importantly, could not ensure a successful outcome. Moreover, even if an agreement was made, the negotiated amount of water transfer was far from adequate to solve Yiwu's water shortage, which can only help it relieve a little from temporary

[103] Yaping Yuan, "The First Water Trading between Cities Performed," *People's Daily,* 07 January 2005.

[104] Shen, *the Studies of the Institutions of Water Rights Transactions, 56.*

[105] Ibid.

[106] China Ministry of Water Resources (MWR), *An Investigation Report on Dongyang-Yiwu Water Rights Trading in Zhejiang Province*, Development Research Center and Economic Regulation Department, Beijing: MWR, 2001, http://www.zjwater.gov.cn/dydoc.asp?docid=774 (accessed September 27, 2008).

[107] Ibid.

[108] Australia China Environment Development Program, *Report on the Development of Water Rights and Trading in the People's Republic of China*, Canberra: Department of Agriculture, Fisheries and Forestry of Australia, 2006, 107.

[109] Shen, *the Studies of the Institutions of Water Rights Transactions*, 56.

droughts. As a result, instead of keeping on the traditional administrative approach, Yiwu was keen to pay Dongyang for more reliable and stable water supply.

An agreement was then made between the two counties in which Dongyang transferred the water use right of 50 million cubice metres of Hengjin Reservoir to Yiwu for RMB 200 million yuan based on the following conditions[110]:

- The transfer of water use rights does not change the original ownership of Hengjin Reservoir and Dongyang is still responsible for the reservoir's operation and maintenance.
- Yiwu pays RMB 0.1 yuan per cubic metre as a comprehensive management fee to Dongyang based on the actual amount of water supplied to it every year. The fee consists of the water resources fee, the cost of the reservoir operation and maintenance, the reservoir depreciation, major repairs, the environmental protection fee, taxes and profits.
- Yiwu is responsible for the planning and investment of the construction of the water diversion project from Hengjin Reservoir to Yiwu. As to the part falling within Dongyang's territory, Yiwu is responsible for the costs while Dongyang is in charge of the construction.

In January 2005, the water diversion project involved in this water trade was completed after 3 years of construction. Hengjin Reservoir began to supply water for Yiwu users. As a result of the fulfilment of the agreement, Dongyang-Yiwu water trade was praised as a win-win deal.[111] The cost of water supply per cubic metre for Dongyang is less than RMB 1.00 yuan.[112] Yiwu, the purchaser, would need to spend RMB 6.00 yuan for per cubic metre water if it built a new reservoir.[113] The price for the agreed water transfer is RMB 4.00 yuan per cubic metre, which means that both parties have obtained the net benefit from the trade.

Nevertheless, this water trade between the two local governments was highly controversial for its legitimacy. According to the water law, water resources are owned by the state and the State Council exercises the ownership of water resources on behalf of the state. Local governments cannot freely dispose of water resources in their jurisdiction but only have the authority in public administration. Some argued that the property right transferred in this trade was not ownership to water resources but the usufruct right. This argument was questionable because the trade was lacking in legal foundation: the relevant laws and regulations did not stipulate the tradable water rights at the time of the trade. The controversy contributed to the delay in the construction of the water diversion project.

Another controversial aspect about the Dongyang-Yiwu trade is related to the specification of the water rights involved in the trade, which led to conflicts between

---

[110] Ibid, 57.

[111] Min Jiang, "Water Rights Trading in China's Water Policy Reform: Current Practices and Future Development," *Proceedings of Water Down Under 2008*, Adelaide: Engineers Australia, 1831–1842.

[112] Shen, *the Studies of the Institutions of Water Rights Transactions*, 65.

[113] Ibid.

the two counties as well as between Dongyang Government and farmers in the irrigation district around Hengjin Reservoir. Dongyang and Yiwu shared the same river for a long time without clearly specifying the initial water allocation quotas. Conflicts would not happen when water resources were abundant and all water users enjoyed sufficient water supply. As water became scarcer with economic and social development, Dongyang as the upstream user was more advantaged in getting water from the river than Yiwu. On the contrary, Yiwu at the lower reaches suffered from water shortage. As the seller of the so-called "water right", Dongyang did not really have legally specified water use rights for the 50 million cubic metres water of Hengjin Reservoir that was sold to Yiwu under the agreement. The trade was made only because Yiwu was willing to accept it. Moreover, Dongyang farmers who used water from Hengjin Reservoir had historic and customary rights to use water for irrigation based on their contribution to the reservoir construction through labour work.[114] Although farmers' water rights were not well specified on a legal basis, their interests should have been taken into consideration under the trade. Unfortunately, they were not consulted about the decision making regarding the essential elements of the trade such as the amount of water for sale. Nor were they properly compensated by either party of the trade. Despite those controversies, the MWR was supportive of the Dongyang-Yiwu trade. Within the context of reform, it is not unusual to see "illegal" events occurring from time to time, which in many cases act as triggers leading to changes in the existing legal system.

Neverthless, the Dongyang-Yiwu water trade was a successful attempt to apply the market-based allocation instrument in China. The benefits of water rights trading were widely recognised even though attitudes towards the trade itself varied. The ineffectiveness of the traditional administrative approach to water allocation was highlighted through the case, which generated a heated debate on the need for clear definition of tradable water property rights in China. The ill-defined water rights in China's present water legal framework was revealed as a major obstacle in developing water trading and the water market. The Dongyang-Yiwu trade took place based on the fact that Yiwu accepted that Dongyang held the water right and took the initiative to propose the purchase. However, an active water market will not appear unless water rights are clearly defined and tradable by law. The other two types of water trading to be discussed in the following subsections occurred in the river basins where water users have relatively well-defined water rights.

4.4.2 Zhangye "Water Notes" Transfer

As one of China's top grain production areas, Zhangye in Gansu Province is the biggest water user of the Hei River Basin (the second largest inland river of China). It is located in the middle stream area of the basin with 91% of the basin's

[114] Ibid, 66–67.

population and 95% of the basin's cultivated land.[115] Its overuse of water from the Hei River has led to severe ecological degradation in the downstream area. Two end lakes of the river, the East Juyanhai Lake and the West Juyanhai Lake dried up respectively in 1961 and 1992, which made Ejina Qi in Inner Mongolia Autonomous Region a major source of Beijing's sand storms.[116]

Since 2000, the State Council has carried out administrative water transfers to restore the ecosystem of the Hei River Basin.[117] The Short-term Restoration Plan of the Hei River Basin approved by the State Council in 2002 confirmed annual water allocation for the downstream area.[118] It means that Zhangye was requested to reduce its water diversion from the river runoff by 580 million cubic metres per year, accounting for 23% of its total annual water use.[119] Within this context, the Zhangye government started the exploration of water rights trading in order to fulfil the challenging task in reducing its water abstraction. In 2002, the MWR approved Zhangye as the first pilot for the "Building a water-saving society" initiative.

The initialisation of water rights was one of the key facilitating factors for water rights trading in Zhangye. Based on the reduction target, water resources allocation schemes were formulated in which the total water use volume was allocated from higher levels to lower levels. At the same time, water use quotas for different using sectors were capped to promote industrial restructuring. At the level of terminal water users, water rights certificates were assigned, which included the detailed information such as the number of family members, the acreage of irrigated farms, types of crops, and the quantity of water supply for each irrigation round.[120]

The relatively well-defined water rights in Zhangye made water trading possible. Irrigators can buy water notes from Water Users' Associations (WUAs) according to their entitled water quantity on water rights certificates. For each irrigation round, water supply is then acquired based on water notes.[121] Those who need more water than their entitled quantity can buy water notes to meet their irrigation needs from other irrigators within the same sublateral canal whose water entitlements are not fully used up. The prices of water notes transfer are negotiated between trading parties with government's guideline which specifies that the prices for water notes transfers must not be three to four times higher than the government-set base price.[122]

The equitability and transparency of water allocation was greatly improved due to the clear definition of water rights in the Zhangye case. Before the water certificate

---

115 Ibid, 80.

116 Ibid, 34–40.

117 For further information about administrative water transfers for ecological emergency, see Sect. 4.2.2.

118 Australia China Environment Development Program, *Report on the Development of Water Rights and Trading in the People's Republic of China, 145.*

119 Shen, *the Studies of the Institutions of Water Rights Transactions*, 81.

120 Ibid, 87.

121 Ibid.

122 Ibid.

system was implemented in the irrigation districts in Zhangye, water charges were prepaid at the beginning of every year and evenly shared by all village irrigators. The balance of water charges may be returned back to irrigators at the end of each year. However, there was no effective monitoring over village officers who were in charge of finances including water tariffs, and irrigators never got the balance back. Farmers had no access to the information of use limits and actual quantity used by each household. As a result, free riding became inevitable. Without clear use limits, the upstream irrigators tend to use more water than their downstream counterparts. Under the new system, irrigators pay for water that is actually used based on specified water entitlements, which to a large extent has reduced free riding and inequity.

Contributing to more than 87% of Zhangye's total water use,[123] agriculture has great potential in water conservation by improving use efficiency. Water notes transfer has effectively encouraged farmers to save water by utilising water-saving technologies and adjusting crop structures. An investigation showed that water use in those pilot irrigation districts in Zhangye reduced by 46 cubic metres per mu[124] while farmers' average incomes increased by RMB 100–150 yuan.[125] Pengzhuang Village in the Hongshuihe Irrigation District provides an example that had the most active water notes trades in Zhangye. There were more than 10,000 cubic metres of water traded between farmers in the village in 2003, which contributed to a 100,000-cubic metre reduction of the total water use.[126]

Compared to the significant impacts of water notes trading on water conservation and total water use control, its effects on economic efficiency remained limited. It was unable to significantly increase economic benefits of water use because water notes only circulate among irrigators within the agricultural sector. Individual farmers only have a small piece of land to cultivate and correspondingly their water entitlements available for trading are also small in number. In addition, although those pilot irrigation districts in Zhangye all operate the system of water rights certificates and provide clear trading rules, it does not necessarily mean that active water rights trading markets exist. Another pilot irrigation district named Liyuanhe, for instance, had no water rights trade reported.[127]

The less active water notes trading might result for various reasons. Unlike the well-developed market economy in Zhejiang Province where the Dongyang-Yiwu case occurred, Zhangye's agriculture-based economy is still heavily influenced by the traditional planned economy. Farmers in those irrigation districts are not familiar with market dealing. On the other hand, because of poor water metering facilities, farmers' water use cannot be accurately monitored. It is not unusual that some

[123] Yuxiu Zhong et al., "Experiences and Implication of Building a Water-Saving Society in Zhangye," *Water Resources Development Research* 7, no. 3 (2003): 45.

[124] 1 mu = 0.0667 hectare.

[125] Australia China Environment Development Program, *Report on the Development of Water Rights and Trading in the People's Republic of China*, 82.

[126] Ibid.

[127] Ibid, 93.

of them keep abstracting groundwater by using their own motor-pumped wells. Farmers therefore lack the interest and willingness to buy water rights through the market since they can get extra water beyond their legitimate entitlement.[128]

To summarise, water notes transfer in Zhangye is the first case of water rights trading between terminal water users in China. Water rights were reallocated among irrigators within one single water-using sector, agriculture. Due to the small scale of China's individual household farming, the main contribution that water notes transfer made was to control the total water use in agriculture rather than moving water resources to higher valued uses. However, the Zhangye water notes transfer case clearly showed the potential of water rights trading in more efficient water allocation. With well-defined water rights available for water users, water rights trading is likely to occur between different water-using sectors. In the Yellow River Basin, where water rights are also relatively well-defined, the paid water rights transfer from agriculture to industry in Ningxia and Inner Mongolia were much larger in scale than Zhangye water notes transfer, which achieved higher economic efficiency. The next subsection reviews the paid water rights transfer from agriculture to industry in the Yellow River Basin.

4.4.3 Paid Water Transfer from Agriculture to Industry in the Yellow River Basin

As introduced in the previous chapter, the Yellow River Basin is one of the most stressful water systems in North China.[129] To protect the fragile ecosystem in the basin, especially in its downstream areas, the government took various measures to control the overuse of water from Yellow River. As early as 1987, the State Council approved the water allocation scheme for Yellow River before the operation of the SNWDP.[130] The scheme defined shares of water use from Yellow River between each provincial region concerned in the basin.

Ningxia and Inner Mongolia are two of the most water-scarce regions in the Yellow River Basin. Ironically, the severe water shortages in these two regions are coupled with huge waste of water use. Agriculture accounted for more than 90% of the total water use with the irrigation utilisation coefficient ratio as low as 0.4.[131] Heavily depending on water supply from the Yellow River, the actual water use by the two regions was in excess of their entitled quotas defined in the State Council's allocation scheme. Ningxia, for instance, used 50% more than its allocated quota

---

[128] Shen, *the Studies of the Institutions of Water Rights Transactions*, 95.

[129] For details about the water stress in the Yellow River Basin, refer to Sect. 2.1.

[130] Xindai An et al., "Prospect of Water Right System Establishment in Yellow River Basin," *China Water Resources* 19 (2007): 66.

[131] Ministry of Water Resources (MWR), *Report on Ningxia and Inner Mongolia's Water Rights Transfer*, Economic Regulation Department, 2004, Beijing: Ministry of Water Resources, http://www.mwr.gov.cn/index/20040416/32859.asp (accessed September 27, 2008).

from 1998 to 2004.[132] While agriculture occupied most of the water available in Ningxia and Inner Mongolia, other growing industries were constrained because of no access to water resources. Although the abundant coal resources help the development of the energy industry, particularly electric power in these two regions, these energy projects needed a large amount of new water supply. For example, 220 million cubic metres of water use had to be increased annually to support the proposed large-scale projects in Inner Mongolia.[133] Between 2002 and 2003, Ningxia and Inner Mongolia submitted several applications for more water abstraction from the Yellow River in order to develop electric power plants. All these applications were rejected by the river basin authority the YRCC,[134] which held back the development of many energy-based projects in Ningxia and Inner Mongolia.

Since water rights became unavailable by new applications for water abstraction from the Yellow River, the energy industry had to look for alternative solutions. As the absolutely dominant water-using sector in Ningxia and Inner Mongolia, agriculture has great potential in water saving by improving its water use efficiency. The challenge, however, is the difficulty in water conservation financing for most of the aged and badly maintained irrigation facilities. Traditionally, irrigation projects heavily rely on investments from both the central and local governments, which remain limited. Within these contexts, the YRCC initiated a pilot programme of water rights transfer from agricultural to industrial use in 2003, to meet increasing water demands without exceeding the total water allocation from the Yellow River. Energy project developers in Ningxia and Inner Mongolia invest in construction of water conservation rehabilitation projects in irrigation districts, and in return, acquire water allocation saved from irrigation by the investments. Water rights transfer from agriculture to industry in the Yellow River Basin was not only effective in water use control, but also in increasing economic benefits of water use. A total of five projects, three in Ningxia and two in Inner Mongolia, were selected as pilots to practise water rights transfer.[135] In 2004, irrigation district management authorities came to agreements of water rights transfer with these fiver pilots.[136] The pilot programme was officially supported by the MWR through its guidelines.[137] To promote the pilots' practices to the entire basin, the YRCC then formulated the Implementing Regulation on Administration of Water Rights Transfer in the Yellow River Basin (Trial) in 2004, which was amended in 2009. According to the 2009

[132] Fengling Tao et al., "The Yellow River's Water Resources Management Based on the Water Rights Market Theory," *Journal of Qinghai University (Nature Science)* 25, no. 1 (2007): 45.

[133] Ibid.

[134] Ibid.

[135] Ibid.

[136] Ibid.

[137] 2004 Guidelines on the Pilot Programme of Water Rights Transfer of the Yellow River's Mainstream in Inner Mongolia and Ningxia (China Ministry of Water Resources), http://www.hwcc.com.cn/newsdisplay/newsdisplay.asp?Id=101638 (accessed September 27, 2008).

regulation, new projects shall obtain water rights through water rights transfer in one of the following circumstances[138]:

- Provincial regions whose water consumption from the Yellow River exceeds its annual share by no more than 5% in two consecutive years.
- Provincial regions that have no remaining water consumption quota within the Yellow River Water Allocation Scheme.
- Administrative regions within prefectures that have no remaining water consumption quota within their detailed provincial total water abstraction control scheme.

Water abstraction right holders of the Yellow River, who have a surplus of water quota in a period of time through water saving projects or changes in water use technologies, can transfer their rights subject to approval by their provincial water management department or the YRCC.[139] The provincial governments are responsible for facilitating water rights trading by formulating their water rights transfer master plan subject to the YRCC's approval.[140] In 2005, Inner Mongolia and Ningxia had their water rights transfer master plans approved by the YRCC, which raised a total of 74 new short-term water rights transfer proposals.[141] More than 600 million cubic metres of agricultural water were proposed to be transferred to other industries such as electric power, chemical technology and coal mining.[142] According to the plans, the two provincial regions would be able to limit their water abstraction from the Yellow River within their entitled quotas by 2015.[143] Farmers benefited from the transfers as their water expenses were largely reduced with rehabilitation of irrigation facilities although they did not receive direct compensation for giving away their water rights. Water rights transfers in the Yellow River Basin were more heavily driven by government than market. The sellers were governmental agencies in charge of water administration or irrigation district management, instead of farmers themselves. As a result, the government-driven practices in water rights transfer tend to take place within local areas such as prefectures due to difficulty in negotiation across different administrative regions.

---

[138] 2004 Trial Implementing Measure on Administration of Water Rights Transfer in the Yellow River Basin (Yellow River Conservancy Commission), http://www.yrcc.gov.cn/ziliao/zcfg/guizhang/200612/t20061222_9390.htm (accessed September 26, 2008), art. 3.

[139] Ibid, art. 5, art. 8.

[140] Ibid, art. 4, art. 9.

[141] Dongliang Yuan et al., "General Plan for Water Right Transfer in Ningxia and Inner Mongolia," *China Water Resources* 19 (2007): 40.

[142] Ibid.

[143] Ibid, 39.

4.5 Major Governance Gaps for Water Rights Trading in China

As discussed in the previous chapters, China is now undergoing a large-scale water reform, of which the most recent policy agenda is to further develop functioning water markets. Using mainly primary data collected from fieldwork in China, this section presents the major constraining factors in China's existing water governance framework that hinder the development of water trading markets. The discussions will help identify legal and policy gaps and inform the future directions of water laws and policies for water rights trading.

The empirical results presented in this section are based on two separate field studies, one conducted in 2006, and the other in 2015. A series of face-to-face semi-structured in-depth interviews were conducted in Zhangye (Gansu Province), Erdos (Inner Mongolia Autonomous Region), and Dongyang (Zhejiang Province) in November 2006.[144] Interviewees included farmers who use irrigation water, other water users, Water Users Associations (WUAs)' staff, and water officials. In 2015, a field study was undertaken to obtain a more up-to-date understanding of China's water governance framework after more than 15 years of substantial efforts in legal, policy, and institutional developments and piloting practices in water rights trading.[145] The 2015 fieldwork included semi-structured interviews[146] (conducted between September and October 2015) and a focus group discussion (held on 27th October 2015 in Beijing) with Chinese water stakeholders across government, industry, and academia. The results from the two field studies provide an in-depth and updated understanding of governance gaps in China's existing water rights trading framework.

4.5.1 Ill-Defined Water Rights

It is widely recognised that well-defined water rights are a prerequisite of a robust water rights market. Although informal water markets can exist in the absence of formal definition of water rights, this type of water trading tends to be limited within a small scale, often within the same water using sector, which largely undermines

---

[144] The fieldwork in 2006 was part of the PhD project of the author entitled "Water rights trading in China: Critical analyses and recommendations for China's water law and policy reform" at Centre for Environmental Law, Division of Law, Macquarie University, Australia.

[145] The fieldwork in 2015 was part of a research project funded by an International Collaborative Research Grant between Victoria University (Australia) and Central University of Finance and Economics (China). The title of the project is "Developing functioning water markets for sustainable water use: Australian experiences and implications for China's water reform", of which the author was the lead chief investigator at Victoria University, Australia.

[146] Most interviews were conducted face-to-face in Beijing except for one telephone interview with an interviewee who was based in Sichuan Province.

water trading's contribution to improving economic efficiency of water use. Clear definitions of water rights not only stimulate active markets by ensuring security of water rights, but also help facilitate larger scale trades between different water using sectors, such as those between agriculture and industries. Empirical analysis focusing on China's pilot water trading cases provided evidence supporting this theoretical argument. It is revealed that the lack of well-defined water rights presents as one of the major constraints to active water rights trading in China.

4.5.1.1 Farmers' Little Sense of Security to Water Rights

In Zhangye, where China's first water rights trading case between terminal water users (water notes transfer) occurred, farmers have relatively well-defined water rights in the form of water rights certificates. Based on their entitled water quantity on water rights certificates, farmers buy water notes from WUAs or irrigation district management agencies and acquire water supply for irrigation according to their water notes each season. Despite its limited contribution to improving economic efficiency of water use, water notes transfer in Zhangye played an important part in conserving water and controlling water use.

However, interviews with individual farmers in Zhangye revealed that they had doubts about the long-term stability of the policy of water notes transfer. The system of water certificates was a part of the efforts in ecological restoration for the Hei River, which was one of many political goals of the government. Because the allocation basis of water certificates was the Short-term Restoration Plan of the Hei River Basin, it gave farmers little confidence in security and value of water certificates as they did not have a strong legal basis either from the Water Law or from the Water Abstraction Permits Regulation. Therefore, as one of interviewees said, they would not be surprised if the government put the initiative off and took the water rights certificates back for nothing.

Famers concerned in the Yellow River water trading case had less confidence in security of their water rights due to the fact that there was no formal definition of farmer level water rights at all. As the water trading practices in the Yellow River Basin were mostly between irrigation districts and industrial developers, regulations only required water rights initialisation at the prefecture level. In some pilot irrigation districts, WUAs were reformed and water rights were further initiated at the level of sublateral canals. The initialisation of water rights at the farmer level is yet to be completed in the basin. As far as they are concerned, those water trading practices were not much different from governmental investments in irrigation rehabilitation. And, if they were asked to tell the difference, they would respond that it is just in who pays for the construction of the projects. Some farmers even thought that they had nothing to do with water rights trading because there were no legally recognised water rights for them.

4.5.1.2 Jeopardised Equity Issues

Lack of clear definition of water rights at the farmer level largely jeopardised the equity issue in water allocation. Farmers' historic or customary water rights were subject to administrative transfer under the traditional government allocation mechanism. Two cases of administrative water allocation: Beijing as paramount priority, and the less favoured agriculture, both made a big compromise on the equity issue. Farmers, like other communities in China, become increasingly aware of the idea of private property rights. While they are willing to sacrifice for political goals like the Olympics, they are consciously advocating for appropriate compensation in the case of water transfer from agriculture to industry.

Farmers also complained about the absence of participatory approaches to engaging them in decision making related to water trading. For example, some farmers concerned in the Dongyang-Yiwu water trade expressed their disappointment in the trade that water rights were permanently transferred without taking into consideration farmers' concern about future water supply. They attributed this to their ill-defined water rights. One interviewee stated that the government did not even need to inform them of the proposed and agreed trade, and potentially their informal water claims could be taken away at any time for any reason. The current practices of water trading are more like quasi-market transactions with government dominating the trades instead of real market players trading freely. As a result, farmers as terminal water users were not highly motivated to conserve water through water rights transfer. The empirical results from fieldwork showed that farmers who compose the largest water consumer in China remain less stimulated to take part in market water transfer because of ill-defined water rights with little security.

4.5.2 Underperformed Water Rights Administration

Moving from a tradition of complete government control over water allocation, China's pilot practices of water rights trading were substantially dominated by government. Governmental authorities played a vital role in water rights trading in China, arguably in an overwhelming manner, more than any other jurisdictions that practise water rights trading in the world. Due to the heavy reliance on public administration, the extent to which water rights trading effectively works as a water allocation mechanism in China is primarily determined by the governance arrangements of governmental authorities as well as their performance in water rights administration.

4.5.2.1 On the Way from Government by Men to Government by Law

To evaluate how the current system of water rights administration performs in China, it may be necessary to first have a look at the Chinese legal ideology that has evolved a unique path through history, independent of other major legal systems in the world.[147] This distinct legal philosophy not only contributes to framing the contemporary water legal system of China but has also had vital and ongoing influence on human behaviour in water rights administration. A brief introduction to such legal philosophy will help understand the context in which water rights administration sits in China.

The Chinese legal ideology involved a combination of influences by two different legal philosophies, namely Government by Men (*Renzhi*) from Confucius and Government by Law (*Fazhi*) from the so-called school of Legalism. The completely opposite stands of these two schools of Chinese philosophy are the result of their different presumptions about human nature.[148] Believing that human nature is fundamentally good, Confucius argued that the foundation of social order was not necessarily the law (*Fa*) but the *Li*, which involved rules of human behaviours and responsibilities for different roles in a web of relationships.[149] If everyone obeyed what they were required to do in accordance with their role in the relationship web, and more importantly, the ruler or officeholder set good examples, the law would be superfluous.

Starting its theories from the idea that human nature was not good as Confucius claimed, the school of Legalism supported the necessity of a fully articulated legal system to which everyone is subject, which is close to the concept of the rule of law prevailing in the western world.[150] Despite the combination of theories rooted in the Confucius and the so-called legalists, the legal thought of Legalism, out of which Government by Law developed, has "remained remote from the minds of the majority of the people, even in modern China".[151] Instead, Government by Men has been the classical and traditional thought of China, which framed its legal system featuring the emphasis on individual obligations rather than rights.[152]

Under such tradition, the administrative power of government tends to be easily abused since there is lack of effective mechanisms to check and supervise its operation. Relationships among the executive, legislative, and judiciary branches of government in China are not effectively governed and regulated by the system of checks and balances present in the western world. In practice, the Chinese judiciary is less independent so that its function as a check on the executive remains minimal. The

[147] Dante A. Caponera, *National and International Water Law and Administration: Selected Writings* (The Hague, London: Kluwer Law International, 2003), 37.

[148] Ibid, 37–39.

[149] Ibid, 31.

[150] Ibid, 39.

[151] Ibid.

[152] Ibid.

legislature stands in a similar situation to the judiciary, failing to ensure equality among the three branches of government.

China is currently committed to transforming its legal system from Government by Men to Government by Law. Enormous efforts have been made to establish the legal framework of public administration that reflects a tendency of regulating and restricting government's administrative actions for a balance between the public administrative power and individual rights. The water rights administration system has developed a range of provisions with regard to the procedures of issuing, modifying and transferring water rights as well as the appeal mechanisms against corresponding administrative decisions.

With these provisions in place, it seems that the current system enables procedures of water rights administration to be transparent to the public and ensures water authorities to be held accountable. However, the full implementation of these provisions in reality is compromised as the traditional legal thought of the people remains. As commented by interviewees, in many cases, water officials, particularly those at lower levels such as prefectures and counties, fail to comply with law. Like their colleagues in other governmental departments, they tend to view themselves privileged over the public. Laws in their eyes are for regulating people rather than themselves.

4.5.2.2 Lack of Transparency and Accountability

Generally speaking, China's governmental authorities have hardly been held accountable for their actions and performances, although the ongoing administrative reform is moving towards a more transparent and accountable government. The lack of accountability was supported by the comments from both water users and water officials interviewed who were involved in water rights administration at the prefecture level. Processing time and information access are two main issues that received most of complaints from water users. For instance, they were not surprised if an application for a new water abstraction permit took half a year to be processed, although the water authorities are required to make the decisions in 45 days from the day of receiving the application. This requirement of 45 days does not include the time needed for public hearings and consultations with other relevant departments, and therefore water authorities could use departmental consultations as a convenient excuse for the delay in processing applications. Water officials were generally not concerned about the time requirements in law, given that they would rarely be accountable for their delays.

Water users were less satisfied about their access to information. According to those interviewed, some water authorities did provide application forms as well as other important information needed for applications, which can be easily accessed from their offices and over the Internet. However, the instructions regarding the eligibility and conditions for water abstraction permit issuance and renewal remain fairly generic and ambiguous to follow. It is difficult for applicants to have access to water plans or water quantity allocation schemes upon which water authorities rely to examine and approve applications. As a result, it fails to ensure transparency that is crucial to provide incentives for water authorities to work more efficiently.

4.5.2.3 High Risk of Rent-Seeking and Corruption

The long-influential tradition of Government by Men leads to high risk of rent-seeking and corruption in public administration of government including water rights administration. Networking, or more commonly known as inter-personal relationships (*Renji Guanxi*), is a powerful resource during the process of public administration in China. A phenomenon called "green light on" (*Kai Lv Deng*) exists in China, which means that a person would be favoured much more than others if they know a government official who may be a family member, a relative, a friend, or even just a stranger but newly introduced by an acquaintance for obtaining the "green light on" through the administration process. Otherwise, it is likely that they will experience intentional difficulties by government officials who often consider themselves privileged over others.

Therefore, almost all water users interviewed stated that they would absolutely use the resource of inter-personal relationships for a smoother and easier process of application as long as there is any available for them. On the other hand, officials of water authorities were generally willing to favour acquaintances since they did not think it was inappropriate, and most importantly, they believed that they may need help from those they helped at some point in the future. Their behaviour, to a large extent, was driven by a hidden rule of Chinese society: inter-personal relationships always deserve to be actively cultivated and maintained for potential future use.

The phenomenon of "green light on" is inevitably associated with high risk of rent-seeking and corruption. If the resource of inter-personal relationships is not available for a person who seeks favour during the application process, or even an undeserved approval for a water abstraction permit, the person may think about buying or exchanging it from water authority officials. Comments from both water users and water officials confirmed the existence of incidents when water officials seek rent during the process of examining and approving water abstraction permits, and issue permits to ineligible applicants.

The empirical results also showed that the ineffective system of check and supervision appears to be the major cause of rent-seeking and corruption in water rights administration. Water authorities, especially those at lower levels, are seldom held accountable for their decisions even though there are some appeal mechanisms present within the current system. In practice, neither administrative reconsiderations nor procedures have been frequently employed by individuals who are concerned about the legitimacy of an administrative decision. Water officials therefore may be able to easily escape from legal liabilities for their misconducts.

The possibility of "buying" water rights from water officials rather than other water users gives new water users a disincentive to enter the water trading market. Accustomed to the tradition of Government by Men, water users interviewed tend to prefer the hidden rule of "green light on" over the new way of getting water rights through market. In fact, a water trade still needs to be approved by water authorities as does a water abstraction permit. The requirement of administration approvals means that water rights trading may not be successful even though two parties have made their agreement. As one interviewee expressed, "why trouble myself since there is another more convenient option?"

4.5.2.4 Potential Conflicts of Interest

Water authorities in China perform more than one function in water resources allocation and management. In some cases, conflicts of interest can easily arise among their various roles in water rights administration. Water authorities may compromise their performance in one function over another based on their own preference.

Conflicts can potentially arise between the jurisdictions in issuance of water abstraction permits and water conservation. Water authorities at local levels heavily rely on their local governments in terms of personnel appointment as well as financial support, although they are also responsible to the MWR at the central level for implementing national water policy. In many cases, local economy is given highest priority over water conservation. Water authorities tend to issue water abstraction permits to applicants whose productive use of water, from an economic perspective, is favoured by the local government.

Water authorities may also be driven to issue water abstraction permits to increase water resources fee collection. Local water authorities are responsible for collecting water resources fees and proportionately forwarding them to the local governments and the central government.[153] Water resources fees are fully incorporated into the fiscal budget of governments and mainly distributed to water-related matters such as water saving, water conservation, water management, and water exploitation. By avoiding the direct control by water authorities over the revenues, this arrangement potentially reduced conflicts of interest. In practice, however, water authorities have close interest in all the above-mentioned water-related matters, which relies on water resources fee collection to bring fiscal support.

Conflicts of interest may also exist between the roles that water infrastructure service providers and water distributors play in water allocation. Water suppliers in China, particularly those providing irrigation water for farmers, are usually a divisional and subordinate unit within water authorities. As a result, a water authority may be examining and approving its own application for water abstraction permits. Commercialisation of water supplies took place in some areas to convert irrigation district authorities from administrative entities to water supply companies. More often than not, water authorities still perform the dual role in water administration and water supply. Like some water officials who participated in the fieldwork

[153] 2006 Regulation on the Administration of Water Abstraction Permits and Water Resources Fee Collection, art. 35.

commented that issues regarding water abstraction permits of irrigation districts can be viewed as internal affairs of water authorities.

4.5.3 Failure in Monitoring and Enforcing Water Rights

Where a water rights system lacks effective monitoring and enforcement arrangements to ensure compliance of water legislation, the value of water rights may be compromised. Water users may ignore the limits of their permits and choose to use water as much as they want if they take no liabilities or penalties for non-compliance. Water rights offer no security when there is no effective monitoring over abstraction of water by water users. Fieldworks undertaken with water stakeholders showed that failure in monitoring and enforcement of water rights presents as a major impediment to water rights trading in China.

4.5.3.1 Hardly Regulated Private Groundwater Pumps

Private groundwater pumps are widely used in China, particularly by farmers and small businesses, as a cheap and reliable water supply. Such uses of private groundwater pumps are often illegal as pump owners hold no permits. Due to the vast amount of private pump users, however, effective regulation is extremely difficult. According to relevant water legislation, a small amount of water abstraction for household uses or for sporadic livestock watering does not need a permit.[154] This provision enables farmers to use their private pumps to extract groundwater for general domestic use. No requirement of administrative approval leaves such water use off regulation and monitoring. As a water authority official commented, water authorities were unable to monitor the volume of water extraction as well as what the water was actually used for by those private pumps.

Even if the illegal use was caught in some cases, private pump owners only needed to pay the water resources fees for the amount of water they extracted rather than getting imposed a penalty for their illegal extraction. This practice not only failed to act as a deterrent to water theft but largely impeded water users' interest in water trading. As an interviewed water user replied: why buy water, since it can come another way cheaper or even free of charge?

4.5.3.2 Lack of Metering Facilities and Effective Monitoring

Historically, China's water use was ineffectively metered except for urban domestic use of tap water. The northern area is much shorter of water availability than the south, and therefore has generally better metering facilities. Nevertheless, it is still

[154] Ibid, art. 4.

a challenge to monitor small businesses, and particularly those small-scale township and village enterprises and farmers. Indeed, area-based water pricing for irrigation remains a common practice in China due to the deficit of metering facilities. According to the data provided by two prefecture water authorities, one in the Yellow River Basin and the other in the Yangtze River Basin, there was no big difference in the installation rate of metering facilities for irrigation water use, with both lower than 2%.

Effective metering is necessary for implementation of total water use control. Regulations explicitly require water abstractors to install metering facilities in accordance with the national technical standard, to ensure their normal operation, and to submit the statistical report of their water abstraction.[155] Abstractors who fail to do so are subject to fines or even revocation of the permits.[156] In reality, water authorities at the local levels experienced tremendous difficulty in implementing such provision. Neither water officials nor water users viewed it as an effective deterrent to non-compliance.

Water abstraction permit holders are also required to report their water abstraction on an annual basis. However, relying mainly on self-monitoring and reporting instead of water authorities doing field inspection, such a supervisory role of water authorities has remained a mere formality. Permit holders hardly worry about sanctions for their abstraction excess of permit limits as they could cover it up by doing "perfect" paper work.

4.5.3.3 Less Enforceable Water Use Control Quotas

As a result of the less regulated private groundwater pumps, lack of metering facilities and effective monitoring, water use control quotas remain hardly enforceable in China. So are water rights. Indeed, it is not unusual to see water taken out of water systems without a permit, or in excess of limit under a permit. Such low level of enforceability in water use control partly explains why water consumption of Inner Mongolia and Ningxia in the Yellow River Basin continually exceeded their water quantity quotas allocated by the 1987 water allocation scheme.

According to the fieldwork data, most irrigation authorities holding water abstraction permits are subordinate agencies of water authorities (river basin authorities, WRBs or WABs) under the current irrigation management framework. Such institutional arrangement would easily raise conflicts of interest, even though they do have internal division of functions. Most interviewees believed that the ineffective enforcement of irrigation water rights was a major attribute to the failure in fulfilling the total water use control in the Yellow River Basin. The water quantity allocation scheme, on the other hand, provided no liabilities or sanctions against the non-complying provinces and regions. Inner Mongolia and Ningxia exceeded their

---

[155] 2006 Regulation on the Administration of Water Abstraction Permits and Water Resources Fee Collection, art. 43.

[156] Ibid, art. 43, art, 53.

water use quotas for several consecutive years, but the YRCC could take no enforcement action against them beyond just rejecting their new abstraction applications.

Using water without permits or in excess of permitted limits is therefore common in China, due to the lack of enforceable water rights down from the individual permit holder level through to the regional level. This undoubtedly reduces the value of water rights and as one interviewee commented, why would a new water user or an existing user with new water demand be willing to purchase water right through the market if it is more available and cheaper somewhere else? Without effective monitoring and enforcement mechanisms in place, the current legal and institutional system in China would be unable to facilitate a dynamic and active water rights trading market.

To summarise, the major constraining factors indicated from the 2006 fieldwork that hinder the development of water trading markets in China include: water rights with little clear specification and security, poorly performed water rights administration, and lack of monitoring and enforcement in water rights. These governance gaps were reaffirmed by the interview and focus group participants in the 2015 fieldwork, although they believed that some improvements were achieved particularly in those piloting places. Comments were made that, while the water reform continues to deepen through the recent nationwide water rights trading piloting practices, the legal and policy developments still lag behind and thus leave those major gaps less addressed in the existing governance framework. Nevertheless, some optimism was present about the expected outcomes of the nationwide piloting programme, for which only time will tell when the programme is to be completed by the end of 2017.

4.6 Concluding Remarks

By reviewing several representative cases, this chapter has examined China's current practices of implementing water allocation instruments. With rapid economic and social development, water resources availability is becoming scarcer in China. The supply-side management has been proved inefficient for sustainable water use, which pushed the Chinese government to move its attention towards the demand-side management options, such as water reallocation and water pricing. Cases reviewed in this chapter have shown some success in the application of those instruments.

Another crucial trend in China's water resources allocation is the implementation of market-based mechanisms. The market transfer of water rights was prohibited by law, and thus most of the necessary institutions remain absent in the current water legal framework. However, several attempts have been made to explore this new approach to water allocation, completely different from China's tradition. It can be found that water rights trading was more likely to first take place at lower levels. On the other hand, the administrative approach tended to be more feasible and effective as the management level went up. However, it is worth noting that, as

Table 4.2 Water allocation instruments used at different levels in China

Levels	Water allocation instruments more likely to be used	Examples
Inter-basin level	Administrative	South-to-north
Within the basin	Administrative	Ecological emergency
Provincial level	Administrative	Beijing-Hebei
Lower local level	Administrative + water rights trading	Dongyang-Yiwu
Terminal water user level	Water rights trading	Zhangye "water notes" transfer

the market-based water allocation mechanisms have been promoted more widely across the country through the new nationwide water rights trading piloting programme, one would see more water rights trading cases at higher management levels such as the regional level and the inter-basin level. While the traditional administrative approach to water allocation has demonstrated its advantages in achieving special political goals, such as water supply security for the 2008 Beijing Olympics, its disadvantages in ensuring water allocation efficiency and equity are also clearly shown. In contrast, China has witnessed the function of the market mechanisms in more efficient water allocation through its pilot practices of water rights trading. Table 4.2 shows the water allocation instruments used at different levels in China.

For a long time, Chinese society and its economy have been developing at the expense of the environment in an unsustainable manner. Such a situation is to be changed. Water rights trading is promoted by the government to not only improve water allocation efficiency but also halt the environmental degradation. Based on the reviews of the water trading cases, it can be concluded that water trading as an allocation instrument has demonstrated its effectiveness in improving economic efficiency as well as controlling unsustainable water use. It also works relatively better than the traditional administrative allocation approach in addressing the equity issue. However, its functionality is undermined by inadequate institutional arrangements of water rights and water rights trading in China. Tradable water rights, for example, have not yet been well defined at terminal water user level. Clear definitions of water rights need to be in place to facilitate a well-functioning and active water trading market. The lack of sufficient legal and institutional arrangements has hindered wider implementation of this market allocation mechanism. The crucial impediments include absent definition of water rights, poor performance of water rights administration, and the failure in monitoring and enforcing water rights. Water users, and in particular farmers, are less encouraged to actively participate in water rights trading.

China's water policy reform aims to establish a national water rights system, within which water rights trading can be implemented more broadly. In 2005, the MWR released its Advices on Water Rights Transfer to encourage more practices. At the same time, the Chinese government has been making a range of efforts in applying water rights trading in the SNWDP. Unlike other public infrastructure constructed and operated by government in the past, this water transfer project is exploring innovative approaches to water allocation. Provincial governments concerned in the project proposed to obtain their water rights based on their shares in the investment of the project.[157] According to the proposal, governments whose demand is in excess of their quotas can acquire more water rights from other governments. At lower levels, water rights trading is also likely to take place between water users in the context of the SNWDP. From this perspective, the SNWDP has potential to become the largest water rights trading market in the world.

To facilitate active and dynamic water trading markets, there is a need to further develop the relevant legal infrastructure, particularly more specified water rights arrangements. Due to the absence of the key preconditions such as initial water rights, it is not surprising to see governments taking responsibilities in coordinating water trades in the early stage of developing the water trading markets. However, in order for this market mechanism to exert its function in efficient water allocation, water rights need to be formally defined to encourage the growth of water trading markets. To give farmers stronger incentives, it is important to make them feel secure about their rights to water. It presents as a major challenge for China to make water rights trading function well in water allocation. How can market forces and government control work smoothly together in the context of China's administration dominated water governance framework? The next chapter addresses the issue of water rights initialisation and arrangements to explore a balance mix of government control and market force in China.

References

An, X., Qing, S., & Chen, Y. (2007). Prospect of water right system establishment in Yellow River basin. *China Water Resources, 19*, 66–69.

Anonymous. (2005). Exploring the water supplier of Beijing, Chicheng County in Hebei: Hardly compensated for its water supply. *New Beijing Newspaper*, 2005. http://www.jxgdw.com/jxgd/news/gnxw/userobject1ai612675.html. Accessed 26 Sept 2008.

Australia China Environment Development Program. (2006). *Report on the Development of Water Rights and Trading in the People's Republic of China*. Canberra: Department of Agriculture, Fisheries and Forestry of Australia.

Barnett, J., Rogers, S., Webber, M., Finlayson, B., & Wang, M. (2015). Transfer project cannot meet China's water needs. *Nature, 527*, 295–297.

Caponera, D. A. (2003). *National and international water law and administration: Selected writings*. The Hague/London: Kluwer Law International.

China Bureau of Planning, Design and Management of the South-to-North Water Diversion. (2003a). Introduction to the overall plan of South to North Water Transfer Project. *China Water Resources, 1*(Part B), 11–13, 18.

China Bureau of Planning, Design and Management of the South-to-North Water Diversion. (2003b). Significant Events of the South-to-North Water Diversion Project. *China Water Resources, 1*(Part B), 86–87.

157 China Ministry of Water Resources, Development Research Centre, "An Introduction to Research on the Construction and Administration System of the South-to-North Project," *China Water Resources* 1B, (2003): 74.

China Ministry of Water Resources. (2001). *An investigation report on Dongyang-Yiwu water rights trading in Zhejiang Province*. Development Research Center and Economic Regulation Department, Beijing: Ministry of Water Resources. http://www.zjwater.gov.cn/dydoc.asp?docid=774. Accessed 27 Sept 2008.

China Ministry of Water Resources. (2004). *Report on Ningxia and Inner Mongolia's water rights transfer*. Economic Regulation Department, Beijing: Ministry of Water Resources. http://www.mwr.gov.cn/index/20040416/32859.asp. Accessed 27 Sept 2008.

China Ministry of Water Resources. (2016). *2015 bulletin of China's water resources*. Beijing: Ministry of Water Resources.

China Ministry of Water Resources, Development Research Centre. (2003). An introduction to research on the construction and administration system of the south-to-north project. *China Water Resources, 1B*, 72–74.

China National Bureau of Statistics (NBS). (2014). *China 2013 Statistical Yearbook*. http://www.stats.gov.cn/tjsj/ndsj/2013/indexch.htm. Accessed 24 Oct 2016.

Cosier, M., & Shen, D. (2009). Urban water management in China. *International Journal of Water Resources Development, 25*(2), 249–268.

Dinar, A., Rosegrant, M. W., & Meinzen-Dick, R. (1997). *Water allocation mechanisms: Principles and examples* (World Bank Policy Research Working Paper, No. 1779). Washington, DC: World Bank, Agriculture and Natural Resources Dept., Sector Policy and Water Resources Division, and, International Food Policy Research Institute.

Food and Agriculture Organisation of the United Nations (FAO). (1999). *Poverty alleviation and food security in Asia: Lessons and challenges*. Bangkok: Food and Agriculture Organisation of the United Nations, Regional Office for Asia and the Pacific.

Gheleta, M. A. (1998). Sustaining the Giant dragon: Rational use and protection of China's water resources in the twenty-first century. *Colorado Journal of International Environmental Law of Policy, 9*, 221–284.

Gleick, P. H. (2000). The changing water paradigm: A look at twenty-first century water resources development. *Water International, 25*(1), 127–138.

Hu, J. (2004). Fight for water rights of the Juma River between Beijing and Hebei. *New Beijing Newspaper*, 2004, http://www.envir.gov.cn/info/2004/2/225855.htm. Accessed 26 Sept 2008.

Jia, Z. Integrated River Basin Management Makes Taitema Lake Green Again. *China Water Resources Newspaper*, 07/07/2006. http://www.mwr.gov.cn/ztpd/2007ztbd/2006ndslhxw/jszgxxwzpj/20070301105420f8f302.aspx. Accessed 26 Sept 2008.

Jiang, M. (2008). *Water rights trading in China's water policy reform: Current practices and future development* (pp. 1831–1842). Proceedings of Water Down Under 2008, Adelaide: Engineers Australia.

Jiang, W., & Tang, Q. (2009). A study on water pricing reform in Beijing. *Journal of Economics of Water Resources, 27*(3), 30–32.

Liao, Y., Gao, Z., Bao, Z., Huang, Q., Feng, G., Xu, D., Cai, J., Han, H., & Wu, W. (2008). *China's water pricing reforms for irrigation: Effectiveness and impact, Comprehensive assessment of water management in agriculture discussion paper 6*. International Water Management Institute: Colombo.

Liu, C. (2000). Environmental issues and the south-to-north water transfer scheme. In R. L. Edmonds (Ed.), *Managing the Chinese environment* (pp. 175–186). Oxford/New York: Oxford University Press.

Lokmar, B., Wang, J., Rozelle, S., Huang, J., & Dawe, D. (2003). *China's agriculture water policy reforms: Increasing investment, resolving conflicts, and revising incentives, Agriculture information bulletin number 782*. Washington, DC: Market and Trade Economics Division, Economic Research Service, U.S. Department of Agriculture.

Ma, D., Lu, X., & Zhang, L. (2007). Current status and management of preexisting pollution via the eastern route of the south-to-north water transfer project. *South-to-North Water Transfers and Water Science & Technology, 5*(4), 29–31.

Magee, D. (2011). Moving the river? China's south-north water transfer project. In S. D. Brunn (Ed.), *Engineering earth: The impacts of magaengineering projects* (pp. 1499–1514). Dordrecht/Heidelberg/London/New York: Springer.

Mao, Z., & Li, Y. (1999, September 1–3). Effect of reform of water charge on water saving in irrigation scheme in China. *The 2nd Inter-Regional Conference on Environment-Water.*

Shen, M. (2006). *The studies of the institutions of water rights transactions—The case studies in China*. Hangzhou: Zhejiang University Press.

Shen, C., & Chen, H. (2010). Cultural heritage management in China: Current practices and problems. In Phyllis Mauch Messenger & G. S. Smith (Ed.), *Cultural heritage management: A global perspective* (pp. 70–81). Gainesville: University Press of Florida.

Sun, F., Yang, Z., & Huang, Z. (2014). Challenges and solutions of urban hydrology in Beijing. *Water Resources Management, 28*, 3377–3389.

Tan, X., Shi, L., Ma, Z., Zhang, X., & Genfa, L. (2015). Institutional analysis of sewage treatment charge based on operating cost of sewage treatment plant – An empirical research of 227 samples in China. *China Environmental Science, 35*(12), 3833–3840.

Tao, X. (2004). The historical development of China's water pricing system. *Zhi Huai, 12*, 16–17.

Tao, F., Ma, B., & Xie, J. (2007). The Yellow River's water resources management based on the water rights market theory. *Journal of Qinghai University (Nature Science), 25*(1), 43–46.

Wang, H., Wang, Z., Pang, J., & Fan, Z. (2011). The issues in the history of administration of water resources fee collection. *Price Monthly, 411*, 42–46.

Webber, M., Crow-Miller, B., & Rogers, S. (2017). The south-to-north water transfer project: Remaking the geography of China. *Regional Studies, 51*(3), 370–382.

Xinhua Net. (2005). *Great impact on cultural relics of the south-to-north project*. Xinhua Net, 2005. http://www.nsbd.gov.cn/zx/mtgz/20040915/200409150024.htm. Accessed 26 Sept 2008.

Xinhua Net. (2007). *Hebei supplies water for Beijing in spite of its consecutive drought for 11 years*. Xinhua Net, 2007. http://www.yznews.com.cn/zhxw/2007-09/28/content_1038853.htm. Accessed 26 Sept 2008.

Yang, J. (2008). Water supply for olympics has led to water conflict. *Nan Feng Chuang*, 10/12/2006. http://www.cnwaternews.com/newscenter/2006/12month/200612102.htm. Accessed 26 Sept 2008.

Yuan, Y. (2005). The First Water Trading between Cities Performed." *People's Daily*, 7 January 2005.

Yuan, D., Wang, T., & Xing, F. (2007). General plan for water right transfer in Ningxia and Inner Mongolia. *China Water Resources, 19*, 38–40.

Zhao, B., Guo, X., & Marinova, D. (2011). To pay more or use less: China's urban water pricing regime and water awareness. In F. Chan, D. Marinova, & S. R. Anderssen (Ed.), *MODSIM2011: 19th international congress on modelling and simulation* (pp. 1724–1730). Perth: The Modelling and Simulation Society of Australia and New Zealand.

Zhong, Y., Liu, H., Yang, N., & Fang, Z. (2003). Experiences and implication of building a water-saving society in Zhangye. *Water Resources Development Research, 7*(3), 45–49.

Zhou, Y. (2002). Study on ecological emergency water transfer for Tarim Lowstream. *China Water Resources, 2*, 68–69.

Zhu, Q., Ren, J., Wang, J., & Sun, F. (2003). Ecological water transfer of China's inland rivers. *China Water Resources, B03*, 49–51.

Legal and Policy Sources

1985 Regulation on Hydraulic Engineering Water Fee Calculation, Collection and Management (China State Council). http://www.hwcc.com.cn/newsdisplay/newdisplay.asp?id=116993. Accessed 27 Sept 2008

1997 Price Law of the People's Republic of China. http://www.npc.gov.cn/englishnpc/Law/2007-12/11/content_1383577.htm. Accessed 22 May 2017.

2002 National Planning Commission's Advices on Agricultural Water Pricing Reform (China National Planning Commission). http://www.hwcc.com.cn/newsdisplay/newsdisplay.asp?Id=38474. Accessed 27 Sept 2008.

2002 Water Law of the People's Republic of China. http://www.npc.gov.cn/englishnpc/Law/2007-12/12/content_1383920.htm. Accessed 21 May 2017.

2003 Regulation on Administration of Water Supply Prices for Water Conservancy Projects (China State Council). http://news.xinhuanet.com/zhengfu/2003-07/22/content_987613.htm. Accessed 27 Sept 2008.

2003 Regulation on the Administration of Urban Water Supply Prices (China State Council). http://www.china5e.com/laws/index2.htm?id=200405090021. Accessed 27 Sept 2008.

2004 Guideline on the Pilot Programme of Water Rights Transfer of the Yellow River's Mainstream in Inner Mongolia and Ningxia (China Ministry of Water Resources). http://www.hwcc.com.cn/newsdisplay/newsdisplay.asp?Id=101638. Accessed 27 Sept 2008.

2004 Trial Implementing Measure on Administration of Water Rights Transfer in the Yellow River Basin (Yellow River Conservancy Commission). http://www.yrcc.gov.cn/ziliao/zcfg/guizhang/200612/t20061222_9390.htm. Accessed 26 Sept 2008.

2006 Regulation on Administration of Water Abstraction Permits and Water Resources Fee Collection (China State Council). http://www.gov.cn/zwgk/2006-03/06/content_220023.htm. Accessed 21 May 2017.

2008 Measures on Management of Water Resources Fee Collection and Use (China Ministry of Finance, National Development and Reform Commission, Ministry of Water Resources). http://www.mof.gov.cn/zhengwuxinxi/caizhengwengao/caizhengbuwengao2008/wengao200811qi/200903/t20090304_118518.html. Accessed 21 May 2017.

2014 Implementing Rules of Residential Water Tiered Pricing for Beijing (Beijing Municipal Commission of Development and Reform). http://zhengwu.beijing.gov.cn/gzdt/gggs/t1351889.htm. Accessed 21 May 2017.

2014 Regulation on Administration of Water Supply and Use for the South-to-North Diversion Project (China State Council). http://en.pkulaw.cn/display.aspx?cgid=219668&lib=law. Accessed 21 May 2017.

Guidelines on Accelerating the Establishment and Improvement of the System of Tiered Water Pricing for Urban Residential Use (China National Development and Reform Commission and the Ministry of Housing and Urban-Rural Development). (2013). http://www.gov.cn/zhengce/2016-05/22/content_5075654.htm. Accessed 22 May 2017.

Minimum Standards for Water Resources Fee Collection by the End of the 12th Five Year Plan (China National Development and Reform Commission, the Ministry of Finance, and the Ministry of Water Resources). (2013). http://www.ndrc.gov.cn/fzgggz/jggl/zcfg/201301/W020130114400736805045.pdf. Accessed 21 May 2017.

State Council General Office's Comments on Promoting the Comprehensive Agricultural Water Pricing Reform (China State Council General Office). (2016). Beijing: China State Council General Office.

Website

The official website of the South-to-North Water Diversion Project. http://www.nsbd.gov.cn. Accessed 21 May 2017.

Chapter 5
Initialising Tradable Water Rights

Abstract This chapter first clarifies the concept of water rights, which is used in various and confusing ways in China. Reflecting the changing philosophy of water use towards sustainability, it then investigates the three-step approach to water rights initialisation: the first step is water planning, which determines how much water can be allocated within a basin or a region as well as among multiple water use sectors. The second step of water quantity allocation distributes quotas among regions and sectors within the basin or the administrative region. It finally comes to the third step of water abstraction permits, which further defines usufruct water rights as the legal basis of tradable water rights. The current legal systems of water planning, quantity allocation, and abstraction permits are examined to understand their deficiencies that may undermine the development of tradable water rights, based on which implications for reform are articulated.

Keywords Water rights initialisation • Philosophy of water use • Water planning • Water quantity allocation • Water abstraction permits

5.1 Application of the Concept of Water Rights

There has been a heated debate on the concept of water rights in the context of China's specific legal system. Indeed, different approaches to the concept of water rights are taken in the literature. While water rights can simply mean all rights related to water in the broadest sense, some would refer to it more specifically. Almost all interpretations agree that water rights refer to the usufruct rights of water resources, the key difference between the approaches lies in whether the concept should include the ownership of water resources.[1] Against those referring water rights to both ownership and usufruct rights, some argue that water rights should not involve ownership since legislation has already established the state property of water resources and prohibited its transfer. The main argument is that the concept of water rights, if including ownership, could cause contradiction and confusion when

[1] Xisheng Huang, *A Study on the Water Rights System* (Beijing: Science Press, 2005), 68–70.

M. Jiang, *Towards Tradable Water Rights*, Global Issues in Water Policy 18,
DOI 10.1007/978-3-319-67087-4_5

talking about "water rights trading".[2] On the other side, this interpretation of water rights is challenged because it makes no distinction between "water rights" and "tradable water rights".[3] In the context of overviewing the framework of water rights initialisation in China, this chapter refers to water rights in its wider application. In other words, the concept of water rights used in this chapter is not only limited within the scope of the usufruct right.

The usufruct right is the main approach to water rights arrangements across many jurisdictions in the world. As Bouviers Law Dictionary defines, usufruct is:

> The right of enjoying a thing, the property of which vested in another, and to draw from the same all the profit, utility and advantage which it may produce, provided it be without altering the substance for the thing. The obligation of not altering the substance of the thing, however, takes place only the case of a complete usufruct.[4]

Unlike a property owner with a full range of rights to use, enjoy and even alienate the property based on their ownership, a usufructuary can use and enjoy the property that belongs to another. Most importantly, the usufruct right does not include the right of alienation although the right itself might be alienated. The usufruct rights over natural resources are substantially different from the traditional usufruct rights in civil law so that some researchers created another title for it, "quasi-property". Generally speaking, quasi-property refers to a group of specific usufruct rights such as mining rights, water rights, fishery rights, and hunting rights.[5] Water rights as a "quasi-property" refer to quasi-property rights of which the holder is entitled to use and receive profits from surface water and groundwater in accordance with law.[6] In addition to water abstraction rights, the group of water rights include water storage rights, disposal rights, navigation rights, bamboo and log rafting rights, etc.

These rights are in essence usufruct rights, but they differ from the traditional definition of a usufruct mainly because of the uniqueness of the objects.[7] For the same reason, even though this group of rights fall into the category of quasi-property rights, each of them might be different from each other based on the unique characteristics of the natural resource on which each quasi-property is established.[8] Under the traditional theory of property rights in civil law, objects of property rights are defined as specific things that always remain stable in quantity and quality, such as

[2] Jianyuan Cui, *Studies on Quasi Property Rights*, 2nd ed. (Beijing: Law Press, 2006), 255. Also see Jianyuan Cui, "Rational Analysis of the Legal Nature of Water Rights," *Journal of Zhengzhou University (Philosophy and Social Science Edition)*37, no, 3 (2004): 18.

[3] Huang, *A Study on the Water Rights System*, 71.

[4] John Bouvier, *A Law Dictionary, Adapted to the Constitution and Laws of the United States*, 1856, http://legal-dictionary.thefreedictionary.com/usufruct (accessed May 7, 2017).

[5] Cui, *Studies on Quasi Property Rights*, 21.

[6] Ibid, 255; Also see Jianyuan Cui, *Property Rights: Growth and Formation* (Beijing: People's University of China Press, 2004), 302.

[7] Cui, *Studies on Quasi Property Rights*, 33.

[8] Ibid, 38.

a piece of land.[9] As the object of water rights, water does not have such specifiability in a strict sense by the traditional definition. Instead, the water over which a water right exists cannot be completely specified due to its fluidity and uncertainty. In this strict sense, some argue that water is non-specifiable and thus water rights are not a type of property right. It is true that the object of water rights is not physically divided from the body of water resources. However, water rights do establish specification in both quantity and quality that is legally recognised and protected, for instance, quantified as a specific amount or a share of water available from a water body.

Another element of quasi-property is that this type of right is not a pure private right but has some characteristics of public rights. Like traditional usufruct rights over properties such as a piece of land or a house, water rights enable right holders to enjoy private interests including taking use of and receiving profits from a certain amount of water. On the other hand, water is also a public good, which means that water rights have to be subject to regulation of public law such as administrative permits, and compared to pure private rights, receive more government intervention in exercising the right.[10] This feature in water rights results from the complex nature of water and water use. Since multiple users have to share it for different purposes, water rights cannot be exclusive like other typical usufruct rights. Furthermore, the right holders receive limitations and restrictions when they exercise their water rights due to the high risk of external effects. In this respect, the conceptualisation of "quasi-property rights" does help understand the unique characteristic of water rights.

In addition to usufructuary rights, there are water supply contracts that are also widely employed as an arrangement of water rights for individual users including farmers. Voluntary water transfers between individual water users on a water supply contract enable flexibility in water reallocation. While a usufruct right is "a right to remove water from the natural environment",[11] a contractual water right is a legal entitlement "to receive a delivery of water through artificial structures".[12] As hydrological engineering develops and more natural water systems are artificially regulated, water supply contracts are increasingly becoming an important form of water rights, particularly for water use in the agricultural sector.

---

[9] Ibid, 36.

[10] Ibid, 63.

[11] Stephen Hodgson, *Modern Water Rights: Theory and Practices* (Rome: Food and Agriculture Organization of the United Nations, 2006), 7.

[12] Ibid.

5.2 The Changing Philosophy of Water Use: Towards Sustainable Water Use

Based on the notion of water use "man can conquer nature", China's water policy had traditionally focused on utilisation and development of water resources rather than water conservation and protection.[13] Indeed, Chinese people in their 5000-year history, had to fight against frequent and disastrous floods numerous times. "The Great Flood" is a vivid example of such traditional philosophy of water use and management. According to Chinese mythology, the Yellow River was prone to flooding, and erupted in a huge flood in the time of Yao, the Emperor.[14] Yao assigned Gun to tame the raging waters, who failed to control the Great Flood after 9 years. Gun's son Yu took his father's place in charge of flood control. Instead of building dikes as his father did, Yu led the people to dredge canals and levees, and eventually solved the flood problems after 13 years of effort (2085–2072 B.C.).[15] Gaining great fame for his achievement in regulating the flood, Yu was selected by Yao as his successor of rulership and was called Yu the Great. Since then, one can easily find many flood fighting stories in the Chinese history.

"Man can conquer nature" has been the predominant philosophy of water use for such a long time in China. It is not only reflected in flood management but also represented by the active utilisation of water resources. By viewing water as nature's gift, Chinese people tend to believe that leaving water in the stream and letting it flow into the sea is a waste. Focusing on utilising water for human activities, water management was oriented towards construction of dams, reservoirs, dikes, and many other water projects.[16] Growing water demands were met by constructing water projects with little attention to the health of the water systems.

As dramatic water demands kept growing along with the social and economic development, water resources were overexploited to an unsustainable degree and caused serious ecological deterioration. The drying-up of the Yellow River was only one example among many others of China's damaged ecosystems. The society suffered from the consequences of its irrational activities. The traditional view of water use was unable to solve water shortage and pollution problems that keep worsening along with the rapidly developing nation. It is now urgent for China to promptly readjust its way of thinking in water management.[17] Aware of the situation in the water crisis, China's water policy reform has initiated a transformation in its approach to water management from project-oriented towards resource-oriented,

---

[13] Shucheng Wang, *Resource-oriented Water Management: Towards Harmonious Coexistence between Man and Nature,* 2nd ed. (Singapore: World Scientific Publishing; Beijing: WaterPower Press, 2006), 37.

[14] One of the Five Emperors. These legendary rulers ruled ancient China between 2850 B.C. and 2205 B.C., the time preceding the Xia Dynasty.

[15] Dante A. Caponera, *National and International Water Law and Administration: Selected Writings* (The Hague, London: Kluwer Law International, 2003), 41.

[16] Wang, *Resource-oriented Water Management,* 3.

[17] Ibid, 34.

Table 5.1 Different purposes of 1988 Water Law and 2002 Water Law

Article 1 of the 1988 Water Law	Article 1 of the 2002 Water Law
This law is formulated for the rational development and utilisation of water resources and the protection of such resources, for the prevention and control of water disasters, and for the full derivation of comprehensive benefits from water resources in order to meet the needs of national economic development and the livelihood of the people.	This law is formulated for the rational development and utilisation of water resources and the conservation and protection of such resources, for the prevention and control of water disasters, and for the sustainable utilisation of water resources in order to meet the needs of national economic and social development.

which implies changes in its notion of water use. The relationship between man and nature is no longer considered absolutely against each other. Rather, resource-oriented water management calls for a new notion of water use: "harmonious coexistence between man and nature".[18]

It is within this context that sustainable use of water resources has been recognised as the purpose of China's water policy. In 2001, the 10th FYP Outline of China's National Economic and Social Development (2001–2005) acknowledged water as one of the strategic resources for development, and sustainable use of water resources was recognised as one of China's development strategies.[19] The 2002 Water Law reaffirmed this principle by amending its first article about the purpose of the law. Table 5.1 illustrates the differences between the 1988 old Water Law and the 2002 new Water Law regarding the purposes of the legislation.

The new philosophy of "harmonious coexistence between man and nature" has been accommodated in the following 11th FYP for Water Resources Development (2006–2010) in 2007, as one of the principles to guide China's water development and reform. It stated that water resources should be rationally developed and utilised in accordance with the carrying capacities of water resources and the water environment, and water development and utilisation should be balanced with ecological and environmental protection.[20] The 12th FYP for Water Resources Development (2011–2015) similarly presented one of the key principles as "insisting on harmony between man and water, and maintaining the health of rivers and lakes".[21]

[18] Ibid, 37.

[19] The 10th Five Year Plan Outline of China's National Economic and Social Development (2001–2005) (China State Council), http://www.people.com.cn/GB/shizheng/16/20010318/419582.html (accessed September 27, 2008), Chapter 1 and Chapter 14.

[20] The 11th Five Year Plan of Water Resources Development of China (China National Development and Reform Commission, Ministry of Water Resources, and Ministry of Construction), 2007, http://www.ndrc.gov.cn/zcfb/zcfbtz/2007tongzhi/t20070607_140130.htm (accessed June 11, 2007), 9.

[21] The 12th Five Year Plan for Water Resources Development of China (China National Development and Reform Commission, Ministry of Water Resources, and Ministry of Housing and Urban-Rural Development), 2012, www.sdpc.gov.cn/fzgggz/fzgh/.../P020160426377473589189.doc (accessed May 22, 2017),11.

The change in the philosophy of water use has also been reflected in increasing attention to demand-side management. Instead of providing new water supplies through technological instruments such as reservoir construction and water diversion to meet growing demands, a range of policy instruments has been applied to control water demand within a limit which would not exacerbate the already severe water shortage and environmental degradation. For example, the 2002 Water Law has established a system of water use control to not only limit the total amount of water use at the macro level but also to set quotas for different water using sectors at the micro level. Regarding the sector level water use quotas, its Article 47 stipulates that,

> The state shall apply a system that combines the control of total amount and the quota management of water use. The competent departments of relevant sectors of the provinces, autonomous regions, and municipalities directly under the State Council shall set sector quotas of water use within their respective administrative areas and submit quotas to the competent departments of water administration, quality supervision and inspection at the corresponding level for examination and approval. Approved quotas shall be promulgated by the people's governments of the provinces, autonomous regions, and municipalities directly under the State Council, and be submitted to the competent departments of water administration, quality supervision and inspection under the State Council for record.[22]
>
> The competent departments of development and planning of the local people's government at or above the county level shall, jointly with the competent departments of water administration at the corresponding level, formulate the annual plan for water use based on water availability in their respective administrative areas determined by their water use quotas, economic and technical conditions, and water allocation schemes. The plan shall control the total amount of the water used annually within their respective administrative areas.[23]

This system is implemented through a number of related systems including water planning, water quantity allocation, and water abstraction permits. As the notion of water use in China transforms towards resource sustainability, major reforms have been undertaken to establish and refine these three systems. Indeed, the reform has attempted to initialise regional water rights and water abstraction rights, which form key elements of the establishment of the national water rights system. The following subsections examine the latest developments of these systems in order to identify governance gaps that need to be addressed in further reform for developing a functioning water rights trading market in China.

[22] 2002 Water Law of the People's Republic of China, http://www.npc.gov.cn/englishnpc/Law/2007-12/12/content_1383920.htm (accessed May 21 2017), art. 47.

[23] Ibid.

5.3 Three Steps of Water Rights Initialisation

5.3.1 Water Planning

Water planning received little attention in the 1988 old Water Law, with only a general article (Article 11) related to planning. With the recognition of its importance in achieving sustainable water use, the regulation on water planning was substantially developed in the 2002 Water Law. Indeed, the 2002 Water Law elaborated on a water planning system with an entire chapter (Chapter 2, Articles 14–19). The chapter addressed various aspects of water planning, such as types of plans, responsible agencies, approval procedures and so on.[24] According to the 2002 Water Law, the water planning framework in China consists of three major parts: national strategic plan for water resources, basin and regional plans, and comprehensive plans and specialised plans.

5.3.1.1 National Strategic Plan for Water Resources

The national strategic water plan reflects the strategic importance of water resources recognised by the Chinese government in its FYP Outline. As Article 14 of the 2002 Water Law states, "the state shall formulate the strategic plan for water resources of the entire country".[25] China's national water resources assessment and utilisation planning was first undertaken in the 1980s. During the following two decades, water planning emphasised utilisation and exploitation of water resources, and paid less attention to water conservation and protection, which resulted in overuse of water and environmental degradation. It was the same period when China also experienced rapid population and economic growth. As water availability and water demand both changed, it became necessary to carry out a new investigation and assessment of China's water resources. In 2002, the MWR and the National Planning Commission jointly issued the Notice to Carry out the Formulation of the National Overall Plan for Water Resources. The mission descriptions of the national overall plan for water resources attached with the Notice listed six key tasks of the planning project[26]:

- To assess water resources and the current situation of utilisation and development of water resources
- To formulate plans for water saving, water protection, and treatment and reuse of wastewater

[24] China 2002 Water Law, chp. 2.

[25] Ibid, art. 47.

[26] *Notice to Carry out the Formulation of the National Overall Plan of Water Resources* (China Ministry of Water Resources and National Planning Commission), 2002, http://www.hwcc.com.cn/newsdisplay/newsdisplay.asp?id=80832 (accessed September 27, 2008).

- To analyse the utilisation potential and carrying capacity of water resources
- To formulate rational allocation scenarios of water resources
- To raise implementation schemes of the arrangements and measures for utilisation, development, control, allocation, conservation, and protection of water resources
- To develop solutions and measures of water management, and establish a water management system consistent with the socialist market economy.

As explained earlier in Chap. 3, although sectoral water management was abandoned by the amended Water Law in 2002, it is still a key characteristic of China's water institutional framework. As a result, various water-related agencies at the central level directly under the State Council are involved in the national overall water planning. Specifically speaking, the central agencies related to water planning include the NDRC,[27] the MWR, the MOHURD,[28] the MEP,[29] the State Commission of Economy and Trade, the Ministry of Land Resources, the Ministry of Agriculture, the State Forestry Administration, and the State Meteorological Administration.

The national overall water planning was carried out based on ten river basin zones.[30] Guided by the MWR technical guidelines, river basin authorities are responsible for formulating detailed rules for overall planning in the river basin and organising provincial level water planning within their basin.[31] Preliminary outcomes of the overall plan for each river basin are further coordinated to produce the national overall water plan. Once the national overall water plan is completed, it acts as the basis for river basin overall plans. Similarly, provincial overall water plans are based on basin plans.[32]

The national water planning was initially expected to complete within 3 years from 2002 to 2005.[33] The three-year timetable was too ambitious to be realistic. The first stage of water investigation and assessment was not completed until late 2004, which was not approved by the State Council until early 2008 after repeated communication and coordination among the participating agencies.[34] The second stage

[27] The National Development and Reform Commission (NDRC) is the former National Planning Commission.

[28] The Ministry of Housing and Urban-Rural Development (MOHURD) is the former Ministry of Construction.

[29] The Ministry of Environmental Protection (MEP) is the former State Environmental Protection Administration.

[30] Including Yangtze, Yellow, Huai, Hai, Songhua, Liao, Pearl, Taihu and south eastern rivers, south western rivers and north western rivers.

[31] Advice on Issues Regarding the Formulation of the National Overall Plan of Water Resources (General Office of the National Planning Commission and General Office of the MWR), 2002, http://www.hwcc.com.cn/newsdisplay/newsdisplay.asp?Id=80937 (accessed September 27, 2008).

[32] Ibid.

[33] MWR and NPC, *Notice to Carry out the Formulation of the National Overall Plan of Water Resources*.

[34] China 2010 National Water Resources Overall Plan (2010–2030) (China State Council), 2010, http://www.sdpc.gov.cn/fzgggz/fzgh/.../P020160426386340300637.doc (accessed May 7, 2017), Forward.

Table 5.2 Main provisions in China's National Strategic Water Plan

Water use total control	The total water consumption for the whole country below 700 billion cubic metres by 2030.
	Define water use total control quota for major river basins and provinces.
Water allocation	Allocation between socio-economic use and in-stream ecological/environmental use.
	Allocation among sectors including domestic, industrial, agricultural, and out-of-stream ecological/environmental use. (the agricultural share will have to reduce down to 57.3%)
	Allocation among different water sources including surface water, groundwater, inter-basin transfer water, and other sources.
	Allocation between rural and urban uses.
	Allocation among provinces within each of the river basin zones.

of water plan formulation also experienced a long delay. The final national water plan was not approved by the State Council until 2010. The delay in the national water planning was firstly due to the fragmented water management. Different sectors use their own assessment standards, which lead to inconsistency and contradiction in data. Coordination between the participating agencies in the planning process was difficult. Secondly, there was a lack of understanding about ecological water, which was for the first time recognised as a new water using sector in national water planning. Both its basic theories and technical calculation methods needed to be further developed. Thirdly, water allocation scenarios involve distributing interests among regions, industries and sectors, and therefore it was difficult to reach consensus among multiple stakeholders. The national water planning was a massive project. It covered 31 provincial areas of China across a range of sectors including water conservancy, agriculture, urban development, and environmental protection.[35] It is an outcome of repeated coordination, cross check, and integration across three levels of administration from the central, through the basin, down to the provincial level.[36] Over 300 organisations participated, from which more than 15,000 people played a direct role in the planning project.[37]

The formulation of the national strategic water plan has significant implications for the development of water trading markets in China. For the first time, sustainability is embedded in China's water plan. As it states in its Foreword, the plan is "a programmatic document for sustainable use of water resources, and provides important basis for development, utilisation, governance, allocation, conservation, protection and management of China's water resources".[38] Table 5.2 summarises the main provisions of the national water plan.[39]

[35] Ibid.

[36] Ibid.

[37] Ibid.

[38] Ibid, 3.

[39] Ibid.

In addition to setting up water use total control quotas for major river basins and provinces, the national water plan calls for the development of water quantity allocation and water abstraction permits total control schemes at lower levels, which will lay the foundations for the establishment of the water rights system.[40]

5.3.1.2 River Basin Plans and Regional Water Plans

Recognising river basins as a key management unit for water resources was a major development of the 2002 Water Law.[41] As its Article 12 stipulates, "the state shall, with respect to water resources, adopt a system that organises the administration by river basins as well as by administrative units".[42] This provision establishes a combined management system for water resources based on national, basin, and local levels. The modification of the water law in river basin plans and regional water plans has reflected this combined approach to water management. According to Article 11 of the 1988 Water Law, "the development and utilisation of water resources as well as the prevention and control of water disasters shall be planned in a unified manner on the basis of river basins or regions".[43] This means that river basin plans and regional water plans were optional. The 2002 Water Law changed this provision by stating in Article 14 that "the development, utilisation, conservation and protection of water resources, and the prevention and control of water disasters, shall be planned in a unified manner one the basis of river basins and regions. Plans are divided into river basin plans and regional plans".[44] Its Article 15 further establishes the relationship between the two types of plans, which is that "the regional plans within a river basin shall be subject to the river basin plan…".[45]

River basin plans include both comprehensive plans and specialised plans. While the comprehensive plans refer to the overall arrangements for development, utilisation, conservation and protection of water resources, and prevention and control of water disasters in river basins, specialised plans focus on specific water-related activities in river basins such as flood prevention, water-logging control, irrigation, navigation, water supply, hydro-electric power generation, bamboo or log rafting, fishery, water resources protection, water and soil conservation, prevention and control of desertification, and water conservation.[46] Similarly, regional plans also include comprehensive plans and specialised plans for provincial regions within a

[40] Ibid.

[41] For further information regarding water management based on river basins and administrative units, refer to Sects. 3.1 and 3.2.

[42] China 2002 Water Law, art. 12.

[43] 1988 Water Law of the People's Republic of China, http://english.mofcom.gov.cn/article/lawsdata/chineselaw/200211/20021100053796.shtml (accessed May 21, 2017), art. 11.

[44] China 2002 Water Law, art. 14.

[45] Ibid.

[46] Ibid.

given river basin. According to the water law, specialised plans shall be subject to comprehensive plans.[47]

The review of comprehensive basin plans for major rivers in China commenced in 2007. These plans were expected to make adjustments consistent with the changing water management approach and water situations. Rather than focusing on utilisation and development of water resources, the review aimed to assure sustainable development of China's river basins.[48] A major part of the review was to limit river utilisation ratios within a sustainable scale. For those rivers where ecosystems are badly deteriorated due to extremely high utilisation ratios, their comprehensive plans need to emphasise ecological rehabilitation.[49]

According to the national overall plan for water resources, the water resources exploitation and utilisation ratios in river basins in northern China are generally high, for example the Hai River Zone has an exploitation and utilisation ratio as high as 134%, followed by the Yellow River Zone with a ratio of 73%.[50] Overexploitation of groundwater in northern China is also severe with an exploitation and utilisation ratio exceeding 100%.[51] To restore the deteriorated surface and groundwater systems, the national water plan has specified water resources availability for each major river basin as the maximum upper control limit for exploitation and utilisation of water resources. In river basins where the actual water consumption has exceeded the water availability, such as the Hai River Zone and the Yellow River Zone, the strictest water conservation measures should be taken to reduce water consumption.[52] Water for ecological conservation and rehabilitation should be increased accordingly in those river basins.[53]

By March 2013, seven major river basin plans[54] had been approved by the State Council. These basin plans have reaffirmed the water conservation and allocation goals set in the national water resources plan, specified control indicators for ecological conservation, and more importantly, defined the "red lines" for total water consumption control, water use efficiency control, and waste disposal control. As discussed in Chap. 2, river basins as main water management units are a key feature of the integrated water resources management. Although China's water law only

---

[47] Ibid.

[48] Advice on the Review of the Comprehensive River Basin Plans (China Ministry of Water Resources), 2007, http://202.123.110.5/zwgk/2007-06/20/content_654854.htm (accessed September 27, 2008).

[49] Yong Jiao, *the Review of the Comprehensive River Basin Plans Guided by the Scientific Concept of Development*, Speech at the Work Meeting of the National Water Planning, 2007, http://www.mwr.gov.cn/ztpd/tszt/lyzhghxbzt/ldjh/200804011648557bc36b.aspx (accessed September 27, 2008).

[50] China 2010 National Water Resources Overall Plan.

[51] Ibid.

[52] Ibid.

[53] Ibid.

[54] The seven river basins include: Yangtze River Basin, Liao River Basin, Yellow River Basin, Huai River Basin, Hai River Basin, Pearl River Basin, Songhua River Basin. In addition, Tai Lake Basin also reviewed its plan, which was approved by the State Council as well.

acknowledges a combined water management approach based on national, basin and local levels, river basin authorities have been given more and clearer mandates in water planning, administration and supervision. The legal status of river basin plans has been accordingly strengthened. As Article 19 of the Water Law stipulates, water project construction in a given river basin must be in conformity with the river basin's comprehensive plan.[55] This provision serves as the fundamental basis for the government to regulate water-related activities and perform water administration within the basin.[56] As the major river basins have renewed their comprehensive plans based on a more sustainable approach to water management, it is expected that overexploitation and deterioration of water systems will be better addressed with implementation of the renewed river basin plans.

5.3.2 *Water Quantity Allocation*

5.3.2.1 Water Quantity Allocation Schemes: Quantifying Water Plans

After water plans have determined the total water availability from a given river basin or other particular area, the determined water availability is further divided between upstream and downstream, among provinces and regions, and among multiple water using sectors. According to Article 44 of the 2002 Water Law, the competent departments of development and planning as well as water administration under the State Council are responsible for the macro-allocation of water resources of the nation.[57] Middle- and long-term plans for water supply and demand at the state, inter-regional and local levels are formulated by considering the present situations of water supply and demand, the national economic and social development plans, river basin plans and regional water plans within the corresponding areas.[58] With river basin plans and middle- and long-term water supply and demand plans in place, the water quantity allocation schemes are formulated on a river basin basis to guide the regulation and storage of the runoff and water quantity allocation.[59]

The development of water quantity allocation schemes has shown disparities among different river basins in China. River basins where serious water conflicts exist, such as the Yellow River and the Hei River, have developed relatively advanced water quantity allocation schemes compared to other rivers. As a result, many river systems in China have been over-used due to the lack of clearly defined water quantity allocation schemes.

From 2004, when pilot practices of water rights trading in several parts of China proved to be effective in reallocating water resources to higher valued uses, the

[55] China 2002 Water Law, art. 19.

[56] Jiao, *the Review of the Comprehensive River Basin Plans*.

[57] China 2002 Water Law, art. 44.

[58] Ibid.

[59] Ibid, art. 45.

Chinese government began to explore water rights initialisation by undertaking pilot practices in the Song-Liao River Basin. As water volumes in most river basins in China have not been clearly allocated to regions at lower levels, it became indispensable to first define the quantitative border of water rights as the starting point of establishing a water rights system. Therefore, most piloting water rights initialisation practices focused on water quantity allocation, which was mainly the formulation of water quantity allocation schemes for river basins, especially those trans-provincial ones.

Two trans-provincial rivers in the Song-Liao River Basin, namely the Huolin River and the Daling River, were selected as pilots to experiment water quantity allocation. With the main task of distributing water volumes among provinces and autonomous regions within the river basin, the pilot practices in the Huolin River Basin and the Daling River Basin made remarkable progress in several aspects of water quantity allocation including its principles, procedures, types and durations, negotiation mechanism, and government volume reserves.[60] Some revolutionary practices occurred in water rights initialisation pilots in the Song-Liao River Basin which later had a significant influence in shaping the MWR's policy document to instruct the nation-wide water quantity allocation. These can be summarised as below:

- The integration of surface water and groundwater into water quantity allocation schemes. Both surface water and groundwater availability are included in the calculation of the total water volumes available in the river basin for allocation. As a result, while different surface water volumes under multiple inflow percentages have been determined for each province or autonomous region as total volume control for surface water use, groundwater exploitation has also been controlled by setting the exploitable groundwater volumes.[61]
- Wetlands are recognised as a water user, which reflects the notion of sustainable water use. Take the Huolin River Basin as an example. As the upper-stream area, Inner Mongolia Autonomous Region has been experiencing a coal and energy boom, which increases water demand. The water needs of the Xiaoghai Wetland in the downstream area cannot be secured unless it can acquire the specified water volume from the water quantity allocation scheme. Fortunately, the scheme did take the ecological water use into account and distributed a volume specifically towards the wetland for the maintenance of its ecosystem.[62]
- Ecological and agriculture water uses are prioritised in allocation schemes to protect disadvantaged water users and facilitate market water rights transfer. Since industrial sectors with increasing water demand are more economically advantaged than ecological water users and farmers, giving priority to wetlands

[60] Australia China Environment Development Program, *Report on the Development of Water Rights and Trading in the Peoples Republic of China* (Canberra, Department of Agriculture, Fisheries and Forestry of Australia, 2006), 182.

[61] Ibid,184.

[62] Ibid,181, 185. Also see Zhongjing Wang et al., "Characteristics of Water Rights System Development in the Huolinhe River Basin," *China Water Resources* 21 (2006): 28.

and agriculture in the stage of water rights initialisation would not only help achieve equity of water allocation but also create incentives for market water rights transfer.[63]

- The government-reserved water volume has been explored in the initial allocation of water quantity as a form of government involvement in water resources management, especially in risk management of water resources. Because of the vital importance of water resources to human survival and community development, government reserves a certain amount of water volume within a river basin prior to the regional allocation as backup to address unpredictable or emergency water demand in the future.[64] While the Huolin River case did not involve a government-reserved water volume due to its inadequate water availability, the allocation in the relatively abundant Daling River did take government reservation into consideration.[65]

5.3.2.2 Legal and Policy Development in Water Quantity Allocation

In the 11th FYP for Water Resources Development, it was highlighted that water quantity allocation schemes for China's major rivers should be developed between 2005 and 2010 as one of the key tasks of the water reform.[66] To implement the national water plan, water quantity allocation will have to be made at all administrative levels. The MWR promulgated the Interim Measures on Water Quantity Allocation in December 2007, which substantially built on the pilot practices of water rights initialisation in the Song-Liao River Basin. The Measures define water quantity allocation in its Article 2 as:

> to distribute the total utilisable amount of water resources or the water quantity available for allocation to administrative regions at various levels, and to specify water quantity quotas for domestic and productive consumption or water abstraction quotas of the administrative regions.[67]

The same article further defines the total utilisable amount of water resources, which reflects the integrated approach to surface water and groundwater:

> The total utilisable amount of water resources includes the utilisable amount of surface water resources and the exploitable amount of groundwater resources deducting the overlapping amount from it.[68]

---

[63] Ibid.

[64] Xinmin Xie et al., "Analysis on the Determination Method of Government-reserved Water Quantities for Initial Water rights Allocation in the Song-Liao Rivers Basin," *Journal of China Institute of Water Resources and Hydropower Research* 4, no. 2 (2006):128.

[65] Australia China Environment Development Program, *Report on the Development of Water Rights and Trading in the Peoples Republic of China*, 183.

[66] The 11th Five Year Plan of Water Resources Development of China, 29.

[67] 2007 Interim Measures on Water Quantity Allocation (China Ministry of Water Resources), http://www.gov.cn/flfg/2007-12/29/content_847012.htm (accessed May 1, 2017), art. 2.

[68] Ibid.

By distributing water volumes from basins to lower administrative levels through provinces and prefectures, the top-down water quantity allocation forms a basis for the implementation of the water abstraction permit system. Water quantity is first distributed among provinces, autonomous regions and municipalities directly under the State Council that share a river basin. These provincial governments then distribute their water quantity quotas down to the next lower administrative regions. The Trial Measures also apply to the water quantity allocation of prefectures down to the next lower administrative regions.[69] Moreover, water use control is carried out on an annual basis through annual water quantity allocation schemes.[70]

In 2011, the MWR released its Notice on Water Quantity Allocation, which included a work plan and technical outline for the formulation of water allocation schemes.[71] It set a goal to develop water allocation schemes for all the major trans-provincial river basins within the 12th FYP period (2011–2015).[72] A group of 25 trans-provincial river basins was listed in the work plan as priority areas for the formulation of water allocation schemes.[73] It was not until August 2016 when most of the water allocation schemes for these 25 river basins were approved by the MWR after a long process of cross-region consultation, review, and amendment.

Based on comprehensive water plans as well as sectoral water use quotas, water quantity allocation clearly defines the total water amount available for multiple users within a river basin and a given administrative unit, serving as a process of quantifying water abstraction rights. In other words, as a determinant of specifying the subjects of water abstraction rights, it is an essential part of water rights initialisation. Water allocation schemes serve as a basis for government to issue water abstraction permits. Without water allocation schemes, the total water use control cannot be effectively achieved.

5.3.3 *Water Abstraction Permits*

5.3.3.1 From Administrative Regulation of Water Abstraction to Legal Basis of Water Rights

In China's traditional water allocation framework, the water abstraction permits system had been applied as an administrative instrument to deliver direct government control over water and water use. The old 1988 Water Law and the 1993 Implementing Measures on the Water Abstraction Permits System provided a

[69] Ibid, art. 3.

[70] Ibid, art. 12.

[71] Notice on Water Quantity Allocation (China Ministry of Water Resources), http://szy.mwr.gov.cn/tztg/tzgg/201405/t20140526_563633.html (accessed May 1, 2017).

[72] Ibid.

[73] Ibid.

framework to supervise water abstraction activities within which water abstraction permits were regulated as a type of administrative licence.

This administrative allocation approach, for a long period of time, emphasised the government's role in water allocation as the administrator of public goods. The legal relationship created by water abstraction permits is different from that between civil parties, which are equal and voluntary. Rather, it tends to feature orders and obedience.[74] As a type of administrative licence, therefore, water abstraction permits do not allow any form of transfers. However, the approach to the administrative regulation over water abstraction has undergone a major change for two main reasons. Firstly, the role of government as the agent to perform the state ownership over water resource has been increasingly recognised, and based on that, the 2002 new Water Law has established the paid use of water resources system. It implies that, rather than providing public services of water supply as a public sector, government should be compensated for the release of water use rights from its ownership. The other reason that led to the change is the need to facilitate market allocation mechanisms. Tradable water property rights are a key element for the development of formal water markets. While water abstraction permits under the tradition of government allocation institution did not assign property rights to water users, the latest developments in China's water policy reform have achieved the separation of water usufructuary rights from water ownership. Water abstraction rights have been explicitly recognised as a property right in the 2007 Property Law. Therefore, the water abstraction permits system has made significant adjustments in consistency with those changes. Water abstraction permits are now arranged as the legal basis of water property rights that are indispensable to nurture the development of formal water markets. As Article 48 of the new Water Law stipulates:

> Entities and individuals that abstract water resources directly from rivers, lakes, or ground waters shall, in accordance with the provisions of the water abstraction permits system and the paid water use system, apply to water administration departments or river basin authorities for a water abstraction license, pay the water resource fees and thus obtain the water abstraction right…[75]

In response to these progresses, the water abstraction permits system experienced some fundamental modification. The 2006 Regulation on Administration of Water Abstraction Permits and Water Resources Fee Collection replaced the 1993 Implementing Measures. In April 2008, the MWR promulgated the Measures on Administration of Water Abstraction Permits to regulate administrative matters in water abstraction permits, including its application, approval, and supervision.

---

[74] Australia China Environment Development Program, *Report on the Development of Water Rights and Trading in the Peoples Republic of China*, 95.

[75] China 2002 Water Law, art. 48.

5.3.3.2 Exemptions from the Water Abstraction Permits System

The water abstraction permits system applies to the abstraction of water directly from rivers, lakes, or groundwaters by utilising water abstraction projects or equipment, such as locks, dams, channels, canals, siphons, pumps, wells and hydroelectric power stations.[76] However, entities or individuals are not required to apply for a water abstraction permit under a range of circumstances. It is necessary to look at these exemptions because they need to be considered when designing arrangements for tradable water rights. The exemptions include[77]:

- Rural collective economic organisations and their members use water in their ponds and reservoirs. The 1988 and 2002 Water Laws took different approaches to property rights regimes over water resources in ponds and reservoirs of rural collective economic organisations. Based on water's attachment with land, the 1988 Water Law entitled the common property of these waters to rural collective economic organisations because they hold the common property to land where these ponds and reservoirs are.[78] In contrast, the state property of water resources was acknowledged as the single property regimes in the 2002 Water Law.[79] Rural collective economic organisations no longer hold the ownership over the waters attached to their common land, but are entitled to usufruct rights. As Article 3 of the 2002 Water Law further states, "water resources in ponds of rural collective economic organisations and in the reservoirs constructed and managed by rural collective economic organisations shall be used by those organisations".[80] Based on their statute usufructuary rights, they are able to abstract water from those ponds and reservoirs with no need to apply for an abstraction permit or pay the water resources fees.
- A small amount of water abstraction for household uses or for sporadic livestock watering is not subject to the water abstraction permits system. This is also a type of statute usufructuary rights mainly aiming to secure domestic water use of farmers. Sporadic livestock watering may go beyond the conventional concept of domestic water, but products from it are essential for the livelihood of poor rural households in China.
- Three other types of temporary emergency water abstraction are exempted from the requirements of water abstraction permits, namely temporary emergency water abstraction and drainage necessary to ensure construction security of

[76] 2006 Regulation on Administration of Water Abstraction Permits and Water Resources Fee Collection (China State Council), http://www.gov.cn/zwgk/2006-03/06/content_220023.htm (accessed May 21, 2017), art. 2.

[77] Ibid, art. 4.

[78] China 1988 Water Law, art. 3. It states that: Water resources shall be owned by the state, that is, by the whole people. The waters of ponds and reservoirs belonging to rural collective economic organizations shall be owned by the collectives.

[79] China 2002 Water Law, art. 3. It states that: Water resources shall be owned by the state. The State Council shall exercise ownership of water resources on behalf of the state.

[80] Ibid.

Table 5.3 Changes in water use priorities in the water law (in the order of prioritised to less prioritised)

The 1988 Water Law (Article 14)	The 2002 Water Law (Article 21)
Domestic use of urban and rural residents	Domestic use of urban and rural residents
Agriculture	Agriculture
Industrial uses	Industrial uses
Navigation	Ecological and environmental uses
	Navigation

Table 5.4 Changes in water use priorities in the water abstraction permits regulation (in the order of prioritised to less prioritised)

The 1993 Implementing Measures (Article 5)	The 2006 Regulation (Article 5)
Domestic use of urban and rural residents	Domestic use of urban and rural residents
Agriculture	Agriculture
Industrial uses	Industrial uses
Navigation	Ecological and environmental use
Environmental protection	Navigation

mines and other underground works; temporary emergency water abstraction necessary to eliminate danger and damage of public security or public interests; and temporary emergency water abstraction necessary to fight droughts for agriculture and to maintain ecosystems and the environment.

5.3.3.3 Priorities of Water Abstraction Permits

How legislation prioritises water abstraction rights is of increasing importance because of the intense competition among multiple water users. In fact, priorities that a type of water right is given by law imply different levels of security of water rights, which can be a key contributing factor for water rights trading. By comparing the two water laws and two regulations on water abstraction permits, it can be seen that the most significant change in water use priority is the environmental water use. Tables 5.3 and 5.4 demonstrate such change.

As shown in the above tables, the old water law did not recognise ecosystems and the environment as a water using sector. By contrast, the new water law not only recognises ecosystems but also gives them a higher priority than navigation. Regulations on water abstraction permits indicated the similar development, within which ecological and environmental water use has now gained higher priority. The different provisions of the two water laws also reflected differently prioritised water uses for water utilisation and development in water short areas. The 1988 Water Law stipulated that "in areas where water sources are insufficient, the scale of the urban area and the development of water intense industrial and agricultural undertakings shall be restricted".[81] By comparison, the 2002 Water Law went further to protect ecosystems by stating that "in arid and semi-arid areas, the development and utilisation of water resources shall take into full consideration the ecological environment's need for water".[82]

Water availability, water use patterns, and economic structures in different regions vary significantly in China. As a result, provincial governments are mandated to determine their own priority list for water abstraction permits in line with their local context on a river basin or region basis.[83] However, most provinces take a similar priority approach to the national water law and regulation.

5.3.3.4 The Volume Control of Water Abstraction

Abstraction is a major type of water use involving consumption of water resources, which means that a large amount of water abstracted from water systems is consumed by human intake or productive inputs. Some portion of the abstracted water may return back to water systems but often with reduced water quality. It is therefore important to get the total volume of water abstraction under control through the permits system, not only to secure adequate inflows to maintain freshwater ecosystems but also to protect them from contamination. In order to do so, both the 2006 Regulation on Administration of Water Abstraction Permits and Water Resources Fee Collection and the 2008 Measures on Administration of Water Abstraction Permits set a cap for water abstraction permits.[84] It is required that the total consumptive volume of water abstraction approved in a river basin shall not exceed the water resources availability for the basin approved by the state.[85] Similarly, the total volume of water abstraction approved in an administrative unit shall not exceed its quota assigned by the river basin authority or the competent department of water resources at the next higher level.

Another restriction for water abstraction permits is the total volume control, which sets a cap particularly for water abstraction in a river basin or an administrative unit in accordance with water allocation schemes or agreements signed by the regions concerned.[86] The 2008 Measures on Administration of Water Abstraction Permits reaffirms this provision with its Article 17:

[81] China 1988 Water Law, art. 14.

[82] China 2002 Water Law, art. 21.

[83] 2006 Regulation on Administration of Water Abstraction Permits and Water Resources Fee Collection, art. 5.

[84] Ibid, art. 7. And 2008 Measures on Administration of Water Abstraction Permits (China Ministry of Water Resources), http://www.gov.cn/flfg/2008-04/17/content_947055.htm (accessed May 21, 2017), art. 4.

[85] As discussed in the previous section on water planning, to restore the deteriorated surface and groundwater systems, the national water plan has specified water resources availability for each major river basin as the maximum upper control limit for exploitation and utilisation of water resources.

[86] 2006 Regulation on Administration of Water Abstraction Permits and Water Resources Fee Collection, art. 15.

> The total volume of water abstraction examined and approved by authorities in charge shall not exceed the total volume control quota of water abstraction permits in the river basin or the administrative unit. New applications shall not be examined and approved in a river basin or administrative unit where the approved total volume has already reached the control quota.[87]

In addition, the approval for water abstraction permits applications is subject to sectoral water use quota. Applicants cannot be approved with a permit where the abstraction volume is larger than the capped water use quota for the sector that they belong to.[88] The total volume of water abstraction permits is also controlled on an annual basis. Authorities in charge of examination and approval of the permits are responsible for decreeing the annual water abstraction plan for the next year to each abstractor who is required to strictly limit their abstraction within the approved amount by the plan.[89]

These volume control arrangements for water abstraction permits have critical implications for water rights trading. Social and economic development keeps generating new claims for water supply. For a long time, China tried hard to meet its growing water demands by unsustainably exploiting its highly stressed waters, which led to over-allocation of many water bodies. Supply-side management failed to alleviate water shortages and solve water use disputes, at the expense of the environment and ecosystems. The water abstraction permits system defined no explicit limitation for new applications in the old 1993 regulation. As a consequence, applicants seeking new water abstraction would have no incentive to initiate water trading but rather follow the old way of applying for a new abstraction permit.[90] It is positive to see that this issue is being addressed. Based on Article 17 of the 2008 Measures mentioned earlier, examination and approval of water abstraction permit applications have to be strictly limited within the total volume control quotas. Water administration departments, river basin authorities and their staff are subject to the provisions of legal responsibilities in the Water Law and the 2006 regulation. If they sign approval documents or issue permits to applicants who are not eligible for statutory conditions,

> They shall be ordered by the administrative or supervisory authorities at the next higher level to correct their misconducts; Where the circumstances are serious, the persons in direct charge and others directly responsible shall be given administrative sanctions in accordance with law; Whereas the case constitutes a crime, they shall be investigated for criminal liabilities in accordance with law.[91]

---

[87] 2008 Measures on Administration of Water Abstraction Permits, art. 17.

[88] 2006 Regulation on Administration of Water Abstraction Permits and Water Resources Fee Collection, art. 15. 2008 Measures on Administration of Water Abstraction Permits, art. 18.

[89] Ibid, art. 38. 2006 Regulation on Administration of Water Abstraction Permits and Water Resources Fee Collection, art. 28.

[90] Interviews with water users in Inner Mongolia on November 7, 2006.

[91] China 2002 Water Law, art. 64. 2006 Regulation on Administration of Water Abstraction Permits and Water Resources Fee Collection, art. 47.

Explicitly restricting the examination and approval of new water abstraction applications in river basins and administrative units of which the control quota has already been exhausted is a necessary policy preparation for the implementation of water rights trading. Prospective water abstractors are more impelled to take the initiative to engage in water rights trading as an alternative to a permit application. In fact, the pilot practices of water rights trading from agriculture to industry in the Yellow River Basin took place as the YRCC rejected the applications from the power plant developers for new water abstraction.[92]

5.3.3.5 The Legal Framework of Water Rights Initialisation

As discussed earlier, water planning, water quantity allocation, and water abstraction permits are the three key components in China's water allocation system through which the separation of usufruct rights from the state ownership to water resources is achieved. Water planning determines how much water can be allocated within a basin or a region and among multiple sectors, and water quantity allocation distributes quotas among regions and sectors within the basin or the administrative region. As a result, water planning and water quantity allocation initialise regional water rights that have multiple levels of rights with provincial, prefecture, and county governments as water rights holders. Clearly specified regional water rights in the form of water allocation quotas make water trading between governments possible. The water abstraction permits system further distributes regional water rights to water users to complete initialisation of user water rights. Water users here might not be terminal water users such as households and farmers, but water suppliers. Clear definition of water rights at the user level is a key element for formal water markets. As shown in the following figure (Fig. 5.1), the water rights initialisation framework in China consists of three levels.

The 2007 Property Law is worth highlighting to better understand this water rights initialisation framework as well as the legal characteristics of water rights. Reflecting the economic relationships of society, a country's property system not only relies on its basic economic and political systems but also responds to its traditional customs and culture. Because of its vital importance in the distribution of interest, the making of the Property Law in China took a slow process a lot longer than other laws. Throughout the 13 years of drafting the law, there were heated debates on its provisions. It was not adopted at the National People's Congress until the eighth submission for examination and approval.[93] The doctrine of "numerus clausus" helps understand why it matters significantly whether a right is stipulated by law or not. The numerus clausus doctrine is a fundamental theory for property law in the civil law system, under which, for security and convenience of transactions, property rights, including their categories and connotations, cannot be created

---

[92] For further details about the paid water transfer from agriculture to industry in the Yellow River Basin, see Sect. 4.4.3.

[93] Ping Jiang, "Ideals and Realities of Property Law," *Forum for Social Sciences* (2007), 59.

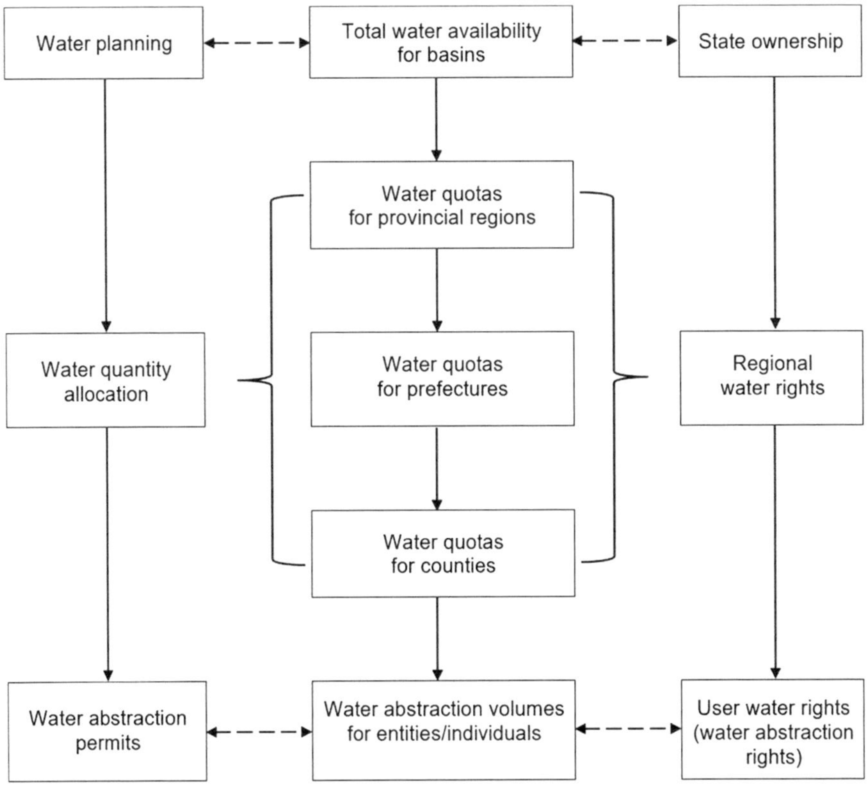

Fig. 5.1 The water rights initialisation framework in China

by transaction parties but by statutory laws.[94] The property law has acknowledged this doctrine in Article 5: "the varieties and contents of property rights shall be prescribed by law".[95] It implies that only those rights that are given a legal title can have complete protection.

For a long time under the traditional of the planned economy, China's waters were more like an open-access resource, even though legislation did provide the state property regime. It mainly resulted from incomplete water allocation planning and poor water metering and monitoring over abstraction permits. As there was lack of clearly set limits on water abstractions for most waters in China, over-issuance of water abstraction permits was common. Moreover, illegal water abstraction and groundwater extraction was not unusual in many areas where water metering facilities were modest. From this perspective, it is not surprising to see "the tragedy of the commons" occurring in many water systems in China. The 2007 Property Law

[94] Ibid, 79.

[95] China 2007 Property Law, art. 5.

includes clear provisions pertaining to state ownership over water resources and several types of water usufructuary rights. The explicit recognition of water use rights as usufruct has significant implications for the establishment of a water rights system and the growth of water rights trading.

The state ownership over water resources is stipulated in Chapters Four and Five in the Property Law. The most relevant articles in these two chapters are as follows[96]:

- The owner of a realty or chattel is entitled to possess, utilise, seek profits from and dispose of the realty or chattel in accordance with law. (Article 39)
- The owner of a realty or chattel is entitled to establish a usufructuary right or real right for security over the realty or chattel. (Article 40)
- As regards a realty or chattel that is exclusively owned by the state as provided by law, its ownership may not be acquired by any entity or individual. (Article 41)
- As regards the properties that shall be owned by the state as provided by law, they shall be in the ownership of the state, that is owned by all the people. The State Council shall exercise the ownership of state-owned properties on behalf of the state; in case there is any otherwise provision in any law, such provision shall prevail. (Article 45)
- Mineral deposits, waters and sea areas shall be in the ownership of the state. (Article 46)
- Unless it is otherwise prescribed by any law, the state practises the system of paid use of natural resources. (Article 119)

According to the law, property owners have a full range of individual property rights that Schlager and Ostrom identified, namely access, withdrawal, management, exclusion and alienation.[97] They are able to either enjoy these rights themselves, or dispose of some or all of these rights. Similarly, as the owner of water resources, the state through the State Council can either take direct control over waters or alienate its property rights to others. However, the right of alienation to waters is restricted due to the state property regime based on the public feature of water resources. This restriction to transfer of water ownership means that the legal arrangements of usufructuary rights are vitally important for optimal resource allocation and utilisation.

Another noteworthy issue is the subject of exercising the state ownership to waters. Statutes recognise the State Council as the only one subject to exercise the state ownership on behalf of the state. Other state organs and public institutions held

[96] 2007 Property Law of the People's Republic of China, http://www.npc.gov.cn/englishnpc/Law/2009-02/20/content_1471118.htm (accessed May 21, 2017).

[97] Edella Schlager and Elinor Ostrom, "Property Rights Regimes and Natural Resources: A Conceptual Analysis," *Land Economics* 68, no. 3 (1992): 250–251. For details, refer to Sect. 2.2.3.

by the state might be able to possess, utilise as well as, in accordance with laws and relevant provisions of the State Council, seek profits from or dispose of state-owned properties that are under their direct control.[98] In reality, however, local governments often dispose of some state-owned properties that they have direct control of, even though there is no specific legal support for alienation. It means that there often exists a de facto ownership right over state properties exercised by local governments although neither laws nor State Council provisions prescribe that local governments have the power to dispose of water resources under their direct control. In fact, the Dongyang-Yiwu water trade between the two county governments was based on such de facto property right, which made it controversial for its legitimacy.[99] The 2016 Provisional Measures on Administration of Water Rights Trading listed regional water rights trading as one of the three types of water rights trading,[100] which for the first time has legitimised that local governments above the county level have the right to dispose of their regional water entitlements thanks to the initialisation of usufructuary rights at the regional levels.

Paid use of water resources and water resources fee collection can also be viewed as an instrument of exercising the state ownership over water resources. Since water resources are owned by the state, any other entities or individuals have to compensate the owner, in form of a water resource fee, for "buying" from the owner the right of possessing and utilising water as well as gaining profits from their possession and utilisation.[101] As discussed in the previous section of this chapter, the water abstraction permit system is transformed from a type of administrative regulation towards legal basis of water usufructuary rights. The water resources fee collection system undertook a similar transformation. Unlike the new 2002 Water Law applying water resources fees to all abstractors, the old 1988 Water Law only required that:

> Any unit that directly extracts groundwater in an urban area shall be charged a water resource fee. As for other water abstraction directly from groundwater or rivers and lakes, people's governments of provinces, autonomous regions or municipalities directly under the State Council shall decide whether water resource fee collection shall apply. Measures for water fee and water resources fee collection shall be stipulated by the State Council.[102]

Therefore, water resources fee collection remained as a provisional mandate until 2006 when the new Regulation on the Administration of Water Abstraction permits and Water Resources Fee Collection was promulgated.

---

[98] China 2007 Property Law, art. 53, art. 54.

[99] For further details about the Dongyang-Yiwu water trade, see Sect. 4.4.1.

[100] 2016 Provisional Measures on Administration of Water Rights Trading (China Ministry of Water Resources), http://www.gov.cn/zhengce/2016-05/22/content_5075679.htm (accessed May 27 2017), art. 3.

[101] Chunmei Mao, "The Implementation of Water Resources Fees in the New Age of China," *China Water Resources* 15 (2007), 8.

[102] China 1988 Water Law, art. 34.

The 1997 Industrial Policy for Water Conservancy for the first time made the explicit connection between paid use of water resources and water resources fee collection through its Article 17:

> The state shall apply for the system of paid use of water resources, and charge water resources when water is directly abstract from groundwater, rivers or lakes. The State Council shall formulate measures on the administration of water resources fee collection and utilisation. The relevant regulations at the provincial level shall temporarily apply before measures of the State Council have been promulgated…[103]

From then on, the water resources fee system was no longer based on government's administrative power to water management, but on the doctrine of paid use of water resources that reflects the state ownership. The 2002 Water Law established this as a fundamental legal system of water management in Article 7:

> The state shall apply the systems of water abstraction permits and paid use of water resources, according to law, except for collective economic organisations and their members when they use water in their own ponds and reservoirs. The competent department of water administration under the State Council shall be responsible for the organisation of the nation-wide implementation of these two systems.[104]

Based on this provision, its Article 48 further stipulates that water resources fee payment together with water abstraction permits are two requirements to obtain water abstraction rights.[105]

As water resources are owned by the state, a usufruct right arrangement is required so that individual water users are able to use water and derive benefit from it. While legal researchers continue to take different approaches to water rights and engage in discussions about "quasi-property rights", water abstraction rights as a type of usufructuary rights is explicitly recognised by the 2007 Property Law. The Property Law provides general rules of usufruct rights including those over natural resources in its Chap. 10. Relevant articles are as below[106]:

- A usufructuary is entitled to possess, utilise and gain profits from a realty or chattel owned by someone else in accordance with law. (Article 117)
- Natural resources owned by the state, owned by the state but used by the collective, or owned by the collective as prescribed by law, may be possessed, utilised and gained profits from by an entity or individual. (Article 118)
- Legal water abstraction rights are under the protection of law. (Article 123)

Therefore, water usufructuary rights are in essence a form of property rights despite the fact that "they gain their existence from an administrative or regulatory procedure" and they "may be subject to restrictions".[107]

---

[103] 1997 Water Conservancy Industrial Policy (China State Council), http://www.cws.net.cn/law/W05.htm (accessed September 27, 2008), art.17.

[104] China 2002 Water Law, art. 7.

[105] Ibid, art. 48.

[106] China 2007 Property Law.

[107] Hodgson, *Modern Water Rights*, 47.

By providing an insightful investigation on the current legal systems as well as their latest developments within the context of the ongoing water reform, it is clear that China is committed to initialising and defining its formal water rights, from water planning through water quantity allocation to water abstraction permits. At the macro level, its water planning and water quantity allocation are underway to limit total water use and to clarify regional water quotas. At the micro level, water abstraction permits are converted from administrative licences to water titles so that the occurrence of water trading markets can be possible. How can China design its tradable water rights within its unique context of traditional approach to administrative allocation? The following section provides a critical analysis of how deficiencies present in the current systems need to be addressed.

5.4 Deficiencies in the Current Legal Systems and Implications for Reform

It can be argued that China has established a preliminary legal and policy framework for water rights initialisation and arrangements, under which usufruct rights over the state-owned water resources are primarily developed by the 2007 Property Law. Regional water rights and user water rights (water abstraction rights) have been initiated through water planning, water quantity allocation and water abstraction permits. This section examines the deficiencies in the current systems as well as the latest progresses, which will provide insights into the implications for the design of tradable water rights.

5.4.1 Ambiguous Regional Water Rights

Despite the reaffirmation of the state ownership of water resources and the State Council as the only one government organisation with the power to exercise the water ownership, the legal nature of regional water rights is still subject to continuing debate. The de facto alienation rights that local governments exercise are at the centre of the debate. Based on the government allocation institution, China's water rights initialisation follows a top-down process from the state ownership exercised by the State Council at the central level, through basins and administrative units at regional levels, then down to water users. Accordingly, it inevitably involves multiple levels of water rights holders. Among these levels, local governments played an important role in implementing water rights trading (the Dongyang-Yiwu case for instance), although the legal legitimacy of regional water rights remained controversial. Such direct government involvement in water rights trading can be viewed as a transitional form, at which the twofold role of local governments as public administration and water right holders is yet to be clearly divided.

In order for the market mechanism to function effectively in water allocation, local governments need to clearly divide their dual parts. Bridging the state ownership and water abstraction rights in between, the regional water rights for local governments were exercised more like their authority in public administration to water resources rather than water rights. According to the 2016 Provisional Measures on Administration of Water Rights Trading, water rights trading refers to water use rights transfer including regional water rights trading.[108] For the first time, regional water rights are legally recognised as a type of water use rights, which presents as a share (quota) granted to a province, prefecture or county of a surface water system or aquifer. Based on regional water rights, the provincial, prefecture or county governments have the right to transfer surplus water within their share to another province, prefecture or county.[109] However, it remains unclarified where the fine line is between government public administration and government as water use right holders. Within its granted share of water use, local governments have the right to allocate that water amongst sub-regions or individuals within the region. What is the legal basis of this allocation right, public administration, or water use rights? This important question is yet to be answered.

5.4.2 *Limited Coverage of Water Abstraction Rights*

In the current systems of water rights initialisation and arrangements, only two specific types of usufruct rights over water resources have been explicitly recognised by the Property Law (Article 123), namely water abstraction rights and rights to use water for breeding or fishery. Water abstraction rights are the main type of tradable water rights in China. Closely associated with competing uses of water including domestic use, irrigation and other productive uses of industries, water abstraction rights result in most water conflicts in China. This is the main driving force why the current water reform made it tradable to promote the market mechanism. However, it is noteworthy that water abstraction rights only refer to very limited water uses. From a horizontal perspective, the current system excludes some essential non-consumptive water uses, such as instream uses of water for the environment. As water scarcity increases, more conflicts arise between consumptive uses and instream uses, particularly the environmental requirements for water. More importantly, the deteriorating freshwater ecosystems in China urge better solutions and higher priority for the environmental water use.

Unfortunately, some key issues related to environmental water rights have barely been addressed in China. Whereas water legislation identifies the environmental need for water and evolves towards more sustainable water use, water rights arrangements still fail to accommodate environmental water sufficiently. The environment as a water user should be recognised by water rights arrangements to balance water's

[108] 2016 Provisional Measures on Administration of Water Rights Trading.

[109] Ibid.

multiple values and prevent the environment from further degradation. From a legal perspective, answers to the following questions need to be explored:

- How can the environmental water rights be accommodated in the system of water abstraction rights that only apply to consumptive water uses?
- Who can be the right holder of an environmental water right?
- What level of priority can the instream environmental use enjoy over other types of water use?
- Should the environmental water rights be made tradable as water abstraction rights?

5.4.3 Less Protected Irrigation Water Rights

The current system based on water abstraction rights is also incomplete from a vertical perspective to cover multiple levels of water users. Only referring to water abstraction directly from a water body, water abstraction rights are granted to a very limited group of water users. Take irrigation as an example. Under the current system, it is usually irrigation district authorities that hold water abstraction rights, and farmers do not have well-defined water rights. According to the Advice on Improving the Establishment of Water Users' Associations (WUAs) of Farmers jointly promulgated by the MWR, the NDRC and the Ministry of Civil Administration in 2005, WUAs can hold water supply contracts while water abstraction rights are conferred to irrigation districts.[110]

To sum up, the current institutional arrangements for irrigation water rights in China include the following:

- Irrigation district authorities (water supply companies in some cases) are primary holders of water rights (water abstraction rights), which have relatively strong legal recognition and protection from the 2002 Water Law and the 2007 Property Law.
- WUAs enter into water supply contracts with irrigation district authorities whose rights to water have not yet been clearly specified in most cases except for those in the water reform pilots. Nor have their contractual rights to water been legally recognised by statute.
- Rights at the farmers' level are generally absent in the current system with exception of reform pilots where farmers hold clearly specified contractual rights to

[110] 2005 Advice on Improving the Establishment of Water Users' Association of Farmers (China Ministry of Water Resources, National Development Reform Commission, and Ministry of Civil Administration), http://law.chinalawinfo.com/newlaw2002/SLC/slc.asp?gid=66344&db=chl&keyword=%E6%94%B9%E9%9D%A9%E5%A7%94%E5%91%98%E4%BC%9A (accessed September 27, 2008).

water in the form of water rights certificates. There is no legal recognition of farmers' water rights certificates.

China's agricultural irrigation has traditionally been administratively managed by irrigation district authorities of local governments. Large scale irrigation districts may have multiple levels of authorities including irrigation district administration, bureaus, stations and branches. In some areas, irrigation district authorities are agencies of local departments in charge of water management, which means that they play a dual role (the one who examines and approves water abstraction permits and the other who holds water abstraction permits).[111] Such institutional arrangement may potentially lead to conflicts of interest. In a few more developed areas, irrigation district authorities have been converted from administrative entities into water supply companies to achieve clear division between the public power as water authorities and property rights based on water abstraction permits. The commercialisation of irrigation districts is an effective means to reduce risk in conflicts of interest.

Within an irrigation district, the management responsibility below the level of sublateral canals is normally given to the village administrators. Water supply is less reliable without clear quantification of water volumes for individual farmers. If shortage occurs, it tends to be upstream irrigators whose water needs get fully satisfied, leaving downstream irrigators with little water supply. However, water charges are still evenly shared by all irrigators regardless of their actual usage, which gives little incentive to improve water use efficiency and conservation.

Within this context, the Chinese government implemented the participatory irrigation management (PIM) to improve irrigation efficiency particularly below the level of sublateral canals. The establishment of WUAs is one of the key instruments among others. Strongly supported by the Chinese government and promoted by international development organisations such as World Bank, the development of WUAs experienced rapid expansion in China. From 1995 when the first WUAs were established in Hubei Province and Hunan Province, the number of WUAs in China reached about 50,000 by 2010.[112]

Some irrigation districts have built WUAs and defined farmers' water rights in the form of water rights certificates. In water reform pilots such as Zhangye, water allocations have been specified for WUAs and their farmers, which played an important role in enabling water trading between farmers (water notes transfer). Yet, a clear specification of water rights at WUA and farmer levels is not in place in many other irrigation districts. Farmers' water rights, represented as water rights certificates, have not yet been legislatively recognised by law. Therefore, farmers' water rights and WUAs still remain unclear in their legal status, which undermines the effectiveness of water rights trading in water reallocation. Research found that WUAs in China failed to deliver for implementation of integrated water resources

[111] Cui, *Studies on Quasi Property Rights*, 298.

[112] Chengbo Xu, "Water Users Associations in Rural China: Thoughts on the Operation Patterns," *China Water Resources* 5 (2010): 21–24.

management at the community level in general, and in particular, did not function effectively in facilitating water rights trading.[113]

In terms of water use priorities for different sectors, although by law agricultural demands stand before industrial and environmental demands just following the domestic demands, in practice, local governments tend to be keen to support demands of "higher value" industrial sectors and view agricultural use as a low priority.[114] Indeed, irrigation water shares continue to shrink in many regions where local governments drive water rights transfer from irrigation districts to industries.

5.4.4 Unclear Boundary for Government Intervention

Under the current institutional framework, water rights as a type of property right are still in the early stage of reform. The private property right remains a fairly new concept for both government and the general public. By looking through water regulations in the Chinese hydraulic civilisation, it can be observed that there was never any private ownership of water.[115] Instead of quoting rights, Chinese water law principles used to just recognise individuals' duties in water-related matters, leading to and enhancing the public good.[116] The tradition was that government took direct supervision of, and responsibility for, water administration. Accordingly, government holds a dominant power in water allocation and largely interferes in major aspects of water rights trading. Major examples of water trading to date have been driven by government instead of emerged as free market. It mainly stems from China's government allocation institution. Such high level of government involvement in water trading practices in China may have its advantages,[117] but in the meantime, overreaching government intervention is likely to undermine the market-based approach to water allocation. How to clearly set the boundary for government intervention in the development and operation of water markets is therefore a critical issue facing China's water reform. In China's context, such boundary is more about how to make water rights secure enough to encourage water users' engagement in water trading. When defining tradable water rights, the major concern should be institutional arrangement for rights certainty and security. From this perspective, the next chapter aims to explore design for well-defined water rights in

[113] Xiaojun Hu et al., "Integrated Water Resources Management and Water Users Associations in the Arid Region of Northwest China: A Case Study of Farmers' Perceptions," *Journal of Environmental Management* 145 (2014): 162–169.

[114] Roger C. Calow et al., "Irrigation Development and Water Rights Reform in China," *International Journal of Water Resources Development* 25, no, 2 (2009): 227–248.

[115] Caponera, *National and International Water Law and Administration*, 31–32.

[116] Ibid, 43.

[117] Robert Speed, "A Comparison of Water Rights Systems in China and Australia," *International Journal of Water Resources Development* 25, no, 2 (2009): 389–405.

China, within which recommendations for future reform will be made based on an analysis of China's specific context as well as reflections on overseas experiences.

References

Australia China Environment Development Program. (2006). *Report on the development of water rights and trading in the Peoples Republic of China*. Canberra: Department of Agriculture, Fisheries and Forestry of Australia.

Bouvier, J. (1856). *A law dictionary, adapted to the constitution and laws of the United States*. http://legal-dictionary.thefreedictionary.com/usufruct. Accessed 7 May 2017.

Calow, R. C., Howardth, S. E., & Wang, J. (2009). Irrigation development and water rights reform in China. *International Journal of Water Resources Development, 25*(2), 227–248.

Caponera, D. A. (2003). *National and international water law and administration: Selected writings*. The Hague/London: Kluwer Law International.

Cui, J. (2004a). *Property rights: Growth and formation*. Beijing: People's University of China Press.

Cui, J. (2004b). Rational analysis of the legal nature of water rights. *Journal of Zhengzhou University (Philosophy and Social Science Edition), 37*(3), 18–20.

Cui, J. (2006). *Studies on quasi property rights* (2nd ed.). Beijing: Law Press.

Hodgson, S. (2006). *Modern water rights: Theory and practices*. Rome: Food and Agriculture Organization of the United Nations.

Hu, X., Xiong, Y., Li, Y., Wang, J., Li, F., Wang, H., & Li, L. (2014). Integrated water resources management and water users associations in the arid region of Northwest China: A case study of farmers' perceptions. *Journal of Environmental Management, 145*, 162–169.

Huang, X. (2005). *A study on the water rights system*. Beijing: Science Press.

Jiang, P. (2007). Ideals and realities of property law. *Forum for Social Sciences*, 59–86.

Mao, C., Fang, G., & Chen, Y. (2007). The implementation of water resources fees in the new age of China. *China Water Resources, 15*, 7–9.

Schlager, E., & Ostrom, E. (1992). Property rights regimes and natural resources: A conceptual analysis. *Land Economics, 68*(3), 250–251.

Speed, R. (2009). A comparison of water rights systems in China and Australia. *International Journal of Water Resources Development, 25*(2), 389–405.

Wang, S. (2006). *Resource-oriented water management: Towards harmonious coexistence between man and nature* (2nd ed.). Singapore/Beijing: World Scientific Publishing/WaterPower Press.

Wang, Z., Zheng, H., Liu, B., Wang, R., & Wang, G. (2006). Characteristics of water rights system development in the Huolinhe River basin. *China Water Resources, 21*, 27–29.

Xie, X., Wang, Z., Wang, J., & Chi, P. (2006). Analysis on the determination method of government-reserved water quantities for initial water rights allocation in the song-Liao Rivers Basin. *Journal of China Institute of Water Resources and Hydropower Research, 4*(2), 128–132.

Xu, C. (2010). Water users associations in rural China: Thoughts on the operation patterns. *China Water Resources, 5*, 21–24.

Legal and Policy Sources

1988 Water Law of the People's Republic of China. http://english.mofcom.gov.cn/article/lawsdata/chineselaw/200211/20021100053796.shtml. Accessed 21 May 2017.

1993 Implementing Measures on the Water Abstraction Permits System (China State Council). http://www.chinawater.net.cn/law/W03.htm. Accessed 28 Sept 2008.
1997 Water Conservancy Industrial Policy (China State Council). http://www.cws.net.cn/law/W05.htm. Accessed 27 Sept 2008.
2002 Water Law of the People's Republic of China. http://www.npc.gov.cn/englishnpc/Law/2007-12/12/content_1383920.htm. Accessed 21 May 2017.
2005 Advice on Improving the Establishment of Water Users' Association of Farmers (China Ministry of Water Resources, National Development Reform Commission, and Ministry of Civil Administration). http://law.chinalawinfo.com/newlaw2002/SLC/slc.asp?gid=66344&db=chl&keyword=%E6%94%B9%E9%9D%A9%E5%A7%94%E5%91%98%E4%BC%9A. Accessed 27 Sept 2008.
2006 Regulation on Administration of Water Abstraction Permits and Water Resources Fee Collection (China State Council). http://www.gov.cn/zwgk/2006-03/06/content_220023.htm. Accessed 21 May 2017.
2007 Interim Measures on Water Quantity Allocation (China Ministry of Water Resources). http://www.gov.cn/flfg/2007-12/29/content_847012.htm. Accessed 1 May 2017.
2007 Property Law of the People's Republic of China. http://www.npc.gov.cn/englishnpc/Law/2009-02/20/content_1471118.htm. Accessed 21 May 2017.
2008 Measures on Administration of Water Abstraction Permits (China Ministry of Water Resources). http://www.gov.cn/flfg/2008-04/17/content_947055.htm. Accessed 21 May 2017.
2016 Provisional Measures on Administration of Water Rights Trading (China Ministry of Water Resources). http://www.gov.cn/zhengce/2016-05/22/content_5075679.htm. Accessed 27 May 2017.
Advice on Issues Regarding the Formulation of the National Overall Plan of Water Resources (General Office of the National Planning Commission and General Office of the MWR). (2002). http://www.hwcc.com.cn/newsdisplay/newsdisplay.asp?Id=80937. Accessed 27 Sept 2008.
Advice on the Review of the Comprehensive River Basin Plans (China Ministry of Water Resources). (2007). http://202.123.110.5/zwgk/2007-06/20/content_654854.htm. Accessed 27 Sept 2008.
China 2010 National Water Resources Overall Plan (2010-2030) (China State Council). (2010). http://www.sdpc.gov.cn/fzgggz/fzgh/.../P020160426386340300637.doc. Accessed 7, May 2017.
Notice on Water Quantity Allocation (China Ministry of Water Resources). (2011). http://szy.mwr.gov.cn/tztg/tzgg/201405/t20140526_563633.html. Accessed 1 May 2017.
Notice to Carry out the Formulation of the National Overall Plan of Water Resources (China Ministry of Water Resources and National Planning Commission). (2002). http://www.hwcc.com.cn/newsdisplay/newsdisplay.asp?id=80832. Accessed 27 Sept 2008.
The 10th Five Year Plan Outline of China's National Economic and Social Development (2001-2005) (China State Council). http://www.people.com.cn/GB/shizheng/16/20010318/419582.html. Accessed 27 Sept 2008.
The 11th Five Year Plan of Water Resources Development of China (China National Development and Reform Commission, Ministry of Water Resources, and Ministry of Construction). (2007). http://www.ndrc.gov.cn/zcfb/zcfbtz/2007tongzhi/t20070607_140130.htm. Accessed 11 June 2007.
The 12th Five Year Plan for Water Resources Development of China (China National Development and Reform Commission, Ministry of Water Resources, and Ministry of Housing and Urban-Rural Development). (2012). www.sdpc.gov.cn/fzgggz/fzgh/.../P020160426377473589189.doc. Accessed 22 May 2017.

Chapter 6
Designing Tradable Water Rights for China

Abstract Well-defined water rights are a central element of a functional water rights market where limited water resources can be reallocated towards higher valued use. The examination of China's water reforms reveals that the ill-defined water rights are one of the major constraints that need to be addressed to make the market mechanism work. This chapter explores how clear and secure tradable water rights can be defined in China's context. Referring to international experiences (e.g. those in Australia and the US) in defining tradable water rights, the chapter elaborates on a number of essential attributes of water rights that affect their security and certainty. These include the origin, duration, predictability and priority of water rights, the level of government interference, transferability and divisibility. A framework of water rights arrangements for China's South-to-North Water Diversion Project (SNWDP) is also proposed to prepare the world's largest water transfer project to become the world's largest water rights trading market.

Keywords Tradable water rights • Origins of water rights • Attributes of water rights • Security of water rights • South-to-North water diversion project

6.1 Origins of Water Rights

To explore the institutional arrangements for tradable water rights, it is necessary to first look at the origins of the rights that have significant implications for the security and transferability of the rights. This section provides a review of legal origins of formal water rights across several overseas jurisdictions including Australia and the United States (US) where water rights trading has been effectively employed in water allocation, in comparison with the institutional arrangements for water rights in China.

M. Jiang, *Towards Tradable Water Rights*, Global Issues in Water Policy 18,
DOI 10.1007/978-3-319-67087-4_6

6.1.1 The Riparian Doctrine

The riparian doctrine is the tradition of water rights in both common law and civil law under which the title to water is tied with the land title. Based on this doctrine, the owner of land bordering a water body acquires rights to the use of the water as that use does not interfere with its use by other riparian landholders.[1] Jurisdictions such as Australia and the US adopted the English common law riparian doctrine.[2] Based on the assumption that there is an abundant supply of water, the riparian doctrine does not provide a basis for water allocation and management.[3] That is why it was the logical choice of the humid eastern states in the US.[4] As originated from the land title, riparian water rights can continue regardless of their use or nonuse.[5] Rather than establishing a priority based on when use was initiated,[6] the riparian doctrine provides for a pro rata sharing of available water supplies with which riparian rights holders have an equal right to use the natural flow of a water course.[7] The principle of "reasonable use" was incorporated as a part of the riparian doctrine which implicitly represents a liability of riparian landholders.[8]

The riparian rights doctrine has also been applied in groundwater. Based on the ownership of the water beneath one's property, the overlying owner has the groundwater right to extract and use it. To an absolute extent, the English law of the riparian doctrine treats the groundwater as a mineral right which even allows the overlying owner to waste the supply without considering the effect on others who share the same resource.[9] In other cases, modifications have been made by courts to restrict the exercise of the right with the requirement of reasonableness, similar to the use of surface water.[10]

The riparian water rights have shown limitations in enabling water use due to the dependence on the adjacent land title. They neither allow water to be moved to higher valued uses nor accommodate the needs of non-adjacent lands for water.[11] By

---

[1] Australia Productivity Commission, *Water Rights Arrangements in Australia and Overseas*, 2003, Melbourne: Productivity Commission, 40.

[2] Michael McKenzie, "Water Rights in NSW: Properly Property?" *Sydney Law Review* 31 (2009): 445.

[3] Jonathan P. Deason et al., "Water Policy in the United States: a Perspective," *Water Policy* 3 (2001):177.

[4] John R. Teerink and Masahiro Nakashima, *Water Allocation, Rights and Pricing: Examples from Japan and the United States*, World Bank Technical Paper, No. WTP 198, 1993, 16.

[5] Ibid.

[6] Ibid.

[7] Ibid. Also see Deason et al., "Water Policy in the United States," 177, and Australia Productivity Commission, *Water Rights Arrangements in Australia and Overseas*, 40.

[8] Australia Productivity Commission, *Water Rights Arrangements in Australia and Overseas*, 40–41.

[9] Ibid, 43.

[10] Teerink and Nakashima, *Water Allocation, Rights and Pricing*, 18.

[11] Deason, et al., "Water Policy in the United States," 177.

tying water to land, the riparian rights can only be exercised in a certain location and for particular purposes so that the flexibility and efficiency of water reallocation are compromised.[12] Therefore, as water stress increasingly worsens, the riparian doctrine has been modified both by statute and judicial actions.[13] For example, most eastern states of the US have moved away from the strict riparian doctrine by enacting new laws to amend or supersede it.[14] Similarly in Australia, the water reform has undertaken to untie water rights from the land title in order to promote water markets.[15]

6.1.2 *Prior Appropriation*

The doctrine of prior appropriation is the dominant law of water allocation in the arid west of the US.[16] It originated in the custom of "fist in time, first in right" that miners had adopted with mineral claims on federal public lands.[17] It was later extended to farmers who needed water for agriculture development as well as other users, including those who privately owned land.[18] The custom later became law by gaining recognition from legislation and court decisions.[19] According to the rule of prior appropriation, "the first person to put water to a beneficial use is entitled to continue that use without interference from those making subsequent uses".[20] Priority of water rights is determined by the time of appropriation, which means that the earlier user (the senior appropriator) has a superior right.[21] In a water shortage, junior water right holders might not be able to access any water while senior water rights in higher priority get fully satisfied first.

Compared to the riparian doctrine that relies on the land title to create water rights, the prior appropriation doctrine establishes water rights based on use of water.[22] During early settlement of the west, most of the land was in the public

[12] Australia Productivity Commission, *Water Rights Arrangements in Australia and Overseas*, 111.

[13] Ibid, 41.

[14] McKenzie, "Water Rights in NSW," 45. Also see Deason et al., "Water Policy in the United States," 178.

[15] Australia National Water Commission (NWC), *Water Markets in Australia: A Short History*, 2011, Canberra: National Water Commission, 29.

[16] Sarah F. Bates et al., *Searching out the Headwaters: Changes and Rediscovery in Western Water Policy*, Washington, D.C., Island Press, 1993, 205.

[17] Ibid, 131.

[18] Ibid, 132–133. Also see Australian Productivity Commission, *Water Rights Arrangements in Australia and Overseas*, 42.

[19] Australia Productivity Commission, *Water Rights Arrangements in Australia and Overseas*, 42. Also see Teerink and Nakashima, *Water Allocation, Rights and Pricing*, 16–17.

[20] Bates et al., *Searching out the Headwaters*, 205.

[21] Australia Productivity Commission, *Water Rights Arrangements in Australia and Overseas*, 42.

[22] Bates et al., *Searching out the Headwaters*, 136.

domain, where the only riparian owner was the United States itself.[23] As a result, water rights established by the prior appropriation doctrine are based on capture and possession (appropriation), which requires water to be diverted out of the stream.[24] As appropriative rights arise from use, they can be permanently in effect as long as right holders continue to divert water and put it to a beneficial use.[25] For the same reason, appropriation rights may be lost if they are not fully used ("use it or lose it").[26]

As a form of private property detached from the land title, a right to use water based on prior appropriation is naturally transferable. Because of the domination of the doctrine in the west, most rivers in the western states of the US were fully or mostly appropriated by the late nineteenth or early twentieth century.[27] To accommodate new water needs of later users, it is logical to allow water rights to be transferred. For instance, new urban and industrial users can purchase existing water rights from farmers. In addition, as new values of water use such as instream uses for environmental and ecological purposes are getting increasingly recognised, the concept of beneficial use is extended beyond the traditional requirement of diversion. Environmental interest groups are able to purchase senior water rights for out-of-stream uses and apply them to instream uses.[28]

6.1.3 Permits

Although water rights have been provided with legal protection as a private property in many jurisdictions adopting either the riparian doctrine or the prior appropriation doctrine, the necessity of government intervention arises from the public nature of water. For this reason, water cannot be fully privatised but has always been subject to public claims.[29] In fact, both the riparian and prior appropriation doctrines are criticised for their limitations in addressing concerns about water's public interests.[30] Within this context, many states in the US have adopted permit systems for water rights as a means of protecting public interests to water. In California, for example, a permit system for surface waters has been established since 1914 under which one must apply for a permit from the State Water Resources Control Board (SWRCB) in order to divert surface water for direct use or store it in a reservoir for

---

[23] Ibid.

[24] Ibid.

[25] Ibid, 147.

[26] Ibid, 142.

[27] Ibid, 148.

[28] Ibid, 157.

[29] Joseph L. Sax, *The Constitution, Property rights and the Future of Water Law*, Boulder, Colo: Natural Resources Law Center, Western Water Policy Project Discussion Series Paper No.2, 1990.

[30] Deason et al., "Water Policy in the United States," 177.

later use.[31] As the state agency to exercise adjudicatory and regulatory functions in water resources, SWRCB is responsible for taking into consideration the protection of instream uses when assessing applications for new water rights.[32]

Under the riparian doctrine, water rights belong to all landowners along the stream and thus are not specified in quantity. In the eastern US states that follow the riparian doctrine, permit systems have also been established to move towards quantification of riparian water rights.[33] By requiring existing riparian water uses to be registered, the permit systems enable riparian rights to achieve same specificity as appropriative water rights.[34] In Florida, for example, failure to obtain a permit during a two-year period may result in a conclusive presumption of abandoning the water use.[35] In this sense, it can be viewed that it is the permit rather than the riparian doctrine from which the water right originates.

The permit systems have also been widely applied by Australian jurisdictions, although water rights may be locally referred to in a variety of terms such as licences, permits, allocations or entitlements.[36] By bringing water resources under statutory control, water legislation was developed which requires riparian water right holders to apply for a licence for their extractions of water other than for livestock watering and domestic purposes.[37] The Australian permit systems often work with water resources planning so that governments have the power to take control of water use for the purpose of environmental protection. Take Queensland as an example. With finalisation of a water resources plan and resource operations plan, water allocations are granted from the conversion of existing entitlements or by applying for new water allocations.[38] A water allocation register is maintained in the Queensland Resource Registry to record water allocations.[39]

Water rights arising from permits can be perpetual or with a specific duration, which make them variable in certainty and security. New water rights issued through permit systems in California are in perpetuity with no time limit,[40] which gives right holders a level of protection similar to that of other private properties. While some states in the US, like California, recognise perpetual water rights, others issue limited tenure rights.[41] In some eastern states, a permit is issued with a fixed duration

[31] Teerink and Nakashima, *Water Allocation, Rights and Pricing*, 21.

[32] Ibid, 22.

[33] Deason et al., "Water Policy in the United States," 179.

[34] Ibid, 179.

[35] Ibid.

[36] Australia Productivity Commission, *Water Rights Arrangements in Australia and Overseas*, 49, 52, 118.

[37] McKenzie, "Water Rights in NSW," 445–446.

[38] Ibid, 52. Also see Australia China Environment Development Program, *Report on the Development of Water Rights and Trading in the Peoples' Republic of China*, Canberra, Department of Agriculture, Fisheries and Forestry of Australia, 2006, 195.

[39] Australia Productivity Commission, *Water Rights Arrangements in Australia and Overseas*, 103.

[40] Teerink and Nakashima, *Water Allocation, Rights and Pricing*, 21.

[41] Bates et al., *Searching out the Headwaters*, 155.

ranging from 10 to 50 years.[42] In Australia, states used to just issue water use licences with limited duration, which could be renewed, amended or cancelled.[43] In Queensland, for example, water allocations have a 10-year period of existence beyond which the state government is able to adjust water allocations through review of water resource plans without compensating right holders.[44] As the water reform has been undertaken to develop water markets, perpetual water access entitlements are established in most states in Australia to encourage investment and maximise water's economic value.[45]

6.1.4 Statute

Besides the common law regimes and the administrative procedures from which water rights can be created, statutory systems can also serve as an origin of water rights. As mentioned earlier, statutes were enacted to legislatively recognise the common law riparian and appropriative water rights in the US. These water rights are still considered to have originated from the riparian doctrine and the prior appropriation doctrine rather than from statutes. Statutory water rights here only refer to those that are directly established by statutes.

Legislation has been enacted not only for the legal recognition of common law regimes but also to facilitate control of water use and allocation and to protect water-dependent ecosystems. Based upon neither public interest nor public policy, the common law does not enable water resources management but only regulates the relationship between water users who have rights of access to the water.[46] In response to the limitation of the common law, statutory systems are applied by states as an instrument to interfere with water resources allocation from the perspective of water's public nature. In other words, legislation enables water to be managed in the public interest.[47]

Generally speaking, those water rights conferred by statutes are naturally non-transferable due to their connection with public values of water resources. For example, statutory water rights normally give stronger protection to some specific uses other than productive uses, including stock and domestic uses, and emergency uses such as protection of life and property against fire. Environmental protection is another concern of statutory water rights. In some US states, federal reserve rights

[42] Ibid.

[43] McKenzie, "Water Rights in NSW," 446.

[44] Australia China Environment Development Program, *Report on the Development of Water Rights and Trading in the People's Republic of China*, 208.

[45] NWC, *Water Markets in Australia*, 51, 56.

[46] Mark Bartley et al., *Trading in Water Rights: Towards a National Legal Framework*, Sydney and Brisbane: Phillips Fox; Queensland University of Technology, 2004, 12.

[47] Ibid.

are conferred on the creation of a nature reserve sometimes by act.[48] Some federal environmental protection legislation also requires environmental water allocations to conserve threatened species or recover endangered species.[49] In New South Wales (NSW), Australia, a statutory management plan is able to establish Environmental Health Water, which is committed exclusively to the ecosystem health of surface water or groundwater.[50]

In other cases, the explicit recognition of water rights in statutes is made to protect specific groups of water users. In some states of the US, the customary rights of Indian tribes have been incorporated within water legislations. In Australia, indigenous rights received little legal recognition until 1992 through the High Court decision in Mabo vs. Queensland (175 CLR 1, 1992) and in response to this landmark decision, the enactment of the Native Title Act (NTA).[51] The NTA recognises Native Title, and sets standards and mechanisms for future claims.[52] Aboriginal interest in water was not incorporated into national policy until 2004 when the National Water Initiative (NWI) was signed by seven members of the Council of Australian Governments (COAG).[53] However, progress on the implementation of the NWI commitments to indigenous access to water at the state level has been slow.[54] As the 2014 Australia's water blueprint national reform assessment commented, most jurisdictions have not yet made specific provision for water access for indigenous people.[55] NSW has established native title rights in its *Water Management Act 2000* to allow native title holders to take and use water without the need for a licence or use approval.[56] Such native title rights are attached to land title record and are not transferable.[57] In many other cases, indigenous water claims in Australia are limited,[58] and the requirement for a specific licence is still in place, for example in Queensland and Northern Territory.[59]

---

[48] Australia Productivity Commission, *Water Rights Arrangements in Australia and Overseas*, 63.

[49] Ibid, 61.

[50] Ibid, 62.

[51] Robert Poirier and Doris Schartmueller, "Indigenous Water Rights in Australia," *The Social Science Journal* 49, no. 3 (2012), 318.

[52] Ibid.

[53] NWI clauses 25 (ix) explicitly recognises the interests of indigenous people. See Australian National Water Commission (NWC), *Third Biennial Assessment of Progress in Implementation of the National Water Initiative*, 2011, http://www.agriculture.gov.au/water/policy/nwi (accessed January 30, 2017).

[54] Ibid.

[55] Australia National Water Commission (NWC), *Australia's Water Blueprint: National Reform Assessment 2014*, http://webarchive.nla.gov.au/gov/20160615062247/http://www.nwc.gov.au/publications/topic/assessments/australias-water-blueprint-national-reform-assessment-2014 (accessed May 18, 2017).

[56] Australia Productivity Commission, *Water Rights Arrangements in Australia and Overseas*, 59.

[57] Ibid.

[58] Poirier and Schartmueller, "Indigenous Water Rights in Australia".

[59] Michael O'Donnell, "The National Water Initiative, Native Title Rights to Water and the Emergent Recognition of Indigenous Specific Commercial Rights to Water in Northern Australia," *The Australasian Journal of Natural Resources Law and Policy* 16, no. 1 (2013): 83–100.

6.1.5 China's Practices

Unlike the US and Australia, where the riparian doctrine was adopted as one of the major legal bases of water rights, China initially did not legally recognise water rights but treated it as a custom in water allocation and especially in solving water disputes among farmers.[60] The institutional arrangements of land ownership in China are vastly different from those of western countries. According to the Constitution and the Property Law of China, land cannot be owned by individuals. Instead, urban land is owned by the state, while rural and suburban land is owned by collectives except for the portions which belong to the state in accordance with law.[61] Because of the state ownership and the collective ownership of land, there is little space for the adoption of the riparian doctrine to establish water rights in China.

However, some statutory water rights have implicitly reflected the essence of the riparian doctrine. As Articles 3 and 7 of the 2002 Water Law state, for instance, collective economic organisations and their members are able to use the water in ponds owned by those organisations and reservoirs constructed and managed by them without being subject to the system of water abstraction permits and the system of paid use of water resources.[62] In addition, according to Article 25, rural collective economic organisations or their members are able to manage and use the water project facilities and the water stored in those facilities that they invest to construct on the collective land owned by the organisation or the land contracted by a member.[63] China's Water Law has also conferred several forms of statutory water rights exempted from the water abstraction permits system. Besides the collective water rights, there are stock and domestic rights and emergency water rights explicitly recognised in China's water legislation.

To better understand China's water rights, it is helpful to make a historic observation to the changes in water abstraction permits since the 1988 Water Law. It was not acknowledged as the main origin of water rights until 2002 when the new Water Law was enacted. The major legal developments are summarised in the following table (Table 6.1).

The State Council enacted the Implementing Measures on the Water Abstraction Permits System in 1993 to put Article 32 of the 1988 Water Law into practice.[64] Its

[60] Jianyuan Cui, *Studies on Quasi Property Rights*, 2nd ed. (Beijing: Law Press, 2006), 299, 302.

[61] 1982 Constitution of the People's Republic of China (2004 Amendments), http://www.npc.gov.cn/englishnpc/Constitution/node_2825.htm (accessed May 21, 2017), art. 10. 2007 Property Law of the People's Republic of China, http://www.npc.gov.cn/englishnpc/Law/2009-02/20/content_1471118.htm (accessed May 21, 2017), art. 47.

[62] 2002 Water Law of the People's Republic of China, http://www.npc.gov.cn/englishnpc/Law/2007-12/12/content_1383920.htm (accessed May 21, 2017), art. 3, art. 7.

[63] Ibid, art. 25.

[64] The state shall put into practice a permit system for abstracting water directly from groundwaters, rivers and lakes. However, it is not required to apply for a permit if water is abstracted for household or for livestock watering, or if a small amount of water is abstracted for other purposes.

Table 6.1 Legal developments of the water abstraction permits system

Old (repealed)	New (current)
The 1988 water law (article 32)	The 2002 water law (article 7, article 48)
The 1993 implementing measures on the water abstraction permits system	The 2006 regulation on administration of water abstraction permits and water resources fee collection
The 1994 provisions on application and approval procedures for water abstraction permits	The 2008 measures on administration of water abstraction permits
The 1995 provisions on water quality management of water abstraction permits	
The 1996 measures on supervision and administration of water abstraction permits	

Article 33 required that any entity or individual, who has already commenced water abstraction before the implementation of the Measures, except those in circumstances listed in Articles 3 and 4,[65] shall register their abstractions to obtain a permit.[66] This provision enabled historic uses of water to be administratively recognised in the form of a permit. At the time, there was generally no effective control over total volume of water available for consumption due to the utilisation-oriented water planning approach and the absence of water sharing plans. Nor was sufficient attention paid to the environment. By recognising the majority of historic abstractions without limiting the issuance of permits within a reasonable cap, many waters of China became over-allocated to an unsustainable level.

Before the 2002 Water Law, water abstraction permits in China tended to be treated like other types of administrative permits rather than as origins of water rights. For example, the duration of a water abstraction permit could not be longer than 5 years.[67] The permit holder had to go to the authority previously in charge of the permit examination and approval for renewal no later than 90 days before the expiry date. Otherwise, the permit would lapse when it expired.[68] In addition, for water abstractors whose permits were issued by the water administrative department under the State Council, river basin authorities, and the water administrative departments at the provincial and prefecture levels or their authorised agencies, and those

[65] Articles 3 and 4 were provisions regarding exemptions from the application for water abstraction permits. 1993 Implementing Measures on the Water Abstraction Permits System (China State Council), http://www.chinawater.net.cn/law/W03.htm (accessed September 28, 2008).

[66] Ibid, art. 33.

[67] 1994 Provisions on Application and Approval Procedures for Water Abstraction Permits (China Ministry of Water Resources), http://www.mwr.gov.cn/zcfg/19940609/13845.asp (accessed June 26, 2008), art. 23.

[68] Ibid.

with urban domestic and industrial uses issued by the water administrative departments at the county level, their conducts were examined on an annual basis.[69]

The implementation of the permit system entered a new era after the enactment of the 2006 Regulation on the Administration of Water Abstraction Permits and Water Resources Fee Collection. Although water quantity allocation schemes and sharing agreements were less available for most of the river basins in China at that time, the MWR required that total volume control indicators for the period from 2006 to 2010 had to be determined through the river basin level[70] to the county level.[71] These indicators coupled with sectoral water use quotas serve as major criteria for examining and approving the amount of water under a permit application. Old permits issued before implementation of the new Regulation had to be replaced within 3 years, through which the amounts of water under the previous permits became subject to an assessment.[72] For instance, the volume of groundwater issued for extraction would be gradually reduced in some areas where groundwater is unsustainably over-extracted.

The water abstraction permit system, which creates water abstraction rights, together with statutory water rights, form the key part of China's water rights arrangements. As mentioned earlier, the permit system in China experienced a transformation from administrative regulation to legal basis for water rights.[73] The assessment of water volumes and replacement of old permits have helped clarify historic claims to water strictly based on the volume control requirements. Recognised as water rights, water abstraction permits have been largely relieved of many administrative restrictions, despite the fact that government still retains a lot of power to interfere with the excise of water rights.

6.1.6 Reflections in the Context of China's Water Reform

The development of water rights origins in overseas jurisdictions has demonstrated a trend in common law regimes moving towards administrative procedures. This is partly because of the increasing concerns of government about protection of public values of water. More importantly, the key driver of such changes is the

---

[69] 1996 Measures on Supervision and Administration of Water Abstraction Permits (China Ministry of Water Resources), http://www.hwcc.com.cn/newsdisplay/NewsDisplay.asp?id=46093 (accessed June 26, 2008), art. 24.

[70] In this sense, river basins refer to those inter-provincial river basins that may include provinces, autonomous regions, or municipalities directly under the State Council.

[71] Notices on the Implementation of the Regulation on Administration of Water Abstraction Permits and Water Resources Fee Collection (China Ministry of Water Resources), 2006, http://www.gdwater.gov.cn/cms/yewuzhuanji/szygl/ggtz/20060703104256.html (accessed September 28, 2008).

[72] Ibid.

[73] For further details about the transformation of water abstraction permits from administrative regulation to legal basis for water rights, refer to Sect. 5.3.3.1.

implementation of water rights trading. For a water market to grow, it is necessary that sellers know exactly what they have available for transfer and, on the other hand, buyers know what they can obtain through a purchase. The administrative permit and registration system has served as a process of making these clear with quantification and specification of water rights.

One should also note the tendency in the enactment of specific legislations to administer and govern water rights. Statutory recognition of water rights not only clarifies the characteristics of water rights but also provides water users stronger protection with clear definition of water users' rights and liabilities. As a civil law country, China does not follow the common law regimes (the riparian doctrine and the prior appropriation doctrine), although there might be some similar ideas embedded in its customs.

Indeed, permits and statutes are the two main sources of formal water rights in China. However, many significant types of water users rights still remain outside of these two water rights regimes, such as those of WUAs and farmers, as well as those for the environment, to name a few. Although in some cases WUAs and farmers can be protected based on their contractual rights to water supply, such protection is far from adequate to motivate their interest in water trading. As the water abstraction permits system only regulates water abstraction directly from water bodies, it is unable to accommodate the contractual rights of WUAs and farmers. The more feasible solution might be legislative recognition of those rights in statute.

While most of the statutory water rights address mainly public interests to water and thus tend to be naturally non-transferable, water rights arising from permits stand at the centre of the institutional arrangements for tradable water rights. To establish an effective system of water rights trading for China, the definition of water rights has to be initiated from the legal infrastructure available in the existing legal framework. Therefore, the next section explores several essential elements in the definition of water rights with sufficient certainty and security, mainly based on the permits system.

6.2 Definition of Tradable Water Rights

6.2.1 Duration

Water rights can be either perpetual or periodic, which affects the certainty and the confidence of the right holder in their future water availability. As far as policy makers are concerned, the duration issue of water rights is "to strike an appropriate balance between the security needed to encourage investments and the need for flexibility as regards future allocations of water".[74] In some western states of the US, such as California and Colorado, water rights are held in perpetuity, and right

[74] Stephen Hodgson, *Modern Water Rights: Theory and Practices* (Rome: Food and Agriculture Organization of the United Nations, 2006), 62.

holders do not need to worry about losing their rights as long as they continue putting the water to beneficial use. In Australian jurisdictions, by comparison, water rights can be perpetual, ongoing or fixed, subject to a 10-year or 15-year review by government.[75] While the NWI set out mechanisms to make water rights perpetual to support water trading markets, the Water Management Act enacted in 2007 by the Commonwealth Government only provides a perpetual share of the water resources available, instead of pure perpetual rights.[76] The review of water rights is often associated with the review of water plans. In NSW, access licences held by irrigation companies in irrigation districts have longer durations. A 20-year term is given to local and major utility licences and a 15-year term to all other types of licences.[77] At the end of the period, water right holders apply for a renewal or extension of their rights. Government, on the other hand, is able to decide whether or not to renew, all or part of a right, and make adjustments of water allocation among users. There is generally no compensation payable for the changes that reduce the amount of water available under a water right beyond the duration.

While the limited tenure of water rights provides government with the flexibility in water resources management,[78] it is not desirable to farmers because of the consequent uncertainty.[79] According to an inquiry by the House of Representatives Standing Committee on Agriculture, Fisheries and Forestry, Australian farmers prefer perpetual water access entitlements, or at least, call for certainty and transparency of the periodic review processes.[80] Aiming to push the Australian water reform in favour of water markets, the 2004 NWI promoted that water access entitlements should be described "as perpetual or open-ended share of the consumptive pool of a specified water resource".[81] As a property right to water, water access entitlements are "a share of the available resource" under the NWI.[82] Intended to deliver investment confidence and security, water access entitlements are defined as a perpetual or long-term right. Australian states have enacted legislation to create secure, perpetual or long-term water access entitlements, including the Australian Capital Territory, NSW, Queensland, South Australia, Tasmania, and Victoria, but not Western Australia or the Northern Territory.[83] The progress in the water entitlement reform in Australia has been appraised for its benefits in facilitating the operation of

[75] Australia Productivity Commission, *Water Rights Arrangements in Australia and Overseas*, 125.

[76] McKenzie, "Water Rights in NSW," 454.

[77] Australia Productivity Commission, *Water Rights Arrangements in Australia and Overseas*, 125.

[78] Ibid, 105.

[79] Australia House of Representatives Standing Committee on Agriculture, Fisheries and Forestry, *Getting Water Right(s)-the Future of Rural Australia*, 2004, Canberra: Commonwealth of Australia, 69–71.

[80] Ibid.

[81] NWC, *Water Markets in Australia*, 58.

[82] NWC, *Australia's Water Blueprint*, 25.

[83] Ibid, 28.

efficient water markets, creating trading opportunities, and reducing transaction costs of trades.[84]

According to the 2006 Regulation on Administration of Water Abstraction Permits and Water Resources Fee Collection, China's water abstraction rights are held in a term between 5 and 10 years.[85] Despite such provision, most water abstraction permits in China are issued with five-year duration. This short duration of water abstraction rights is likely to make right holders less confident in their ability to take water in the future and thus give them less incentive to make investments. Furthermore, how the term range between 5 and 10 years is determined for a specific water use remains unclear in the current legislation.

At the end of the duration, water abstractors need to submit applications for renewal. The 2006 Regulation does not provide a set of rules on how the renewal will be operated to enable right holders to have a reasonable expectation on the renewal results. The Measures on Administration of Water Abstraction Permits enacted in 2008 include provisions regarding the assessment for water abstraction permit renewal. According to the 2008 Measures, authorities in charge of examining and approving water abstraction permits must take the following factors into consideration to determine whether a water abstraction permit should be renewed[86]:

- The volume of water approved in the previous permit
- The volume of water actually abstracted by the permit holder
- The water saving level
- The quality of return flows
- The average water use level of the sector to which the water abstractor belongs
- The situation of water supply and demand in the local region

In the context of the implementation of the strictest water resources management system, it is expected that many water abstraction permits may be discontinued when they reach their expiry date.

In short, water users in China have more worries about the security and certainty of their water rights than their counterparts overseas due to the short duration and uncertainty in the administrative renewal of water rights. Short-term water rights bring right holders lesser confidence in the status and tenure of their right, which discourages investment and trading. To stimulate trading in water markets, China may have to consider extending the duration of its water abstraction permits. Moreover, by learning from the Australian experiences, a system of differentiated (longer than the present) durations for different types of water use may be introduced

[84] Ibid.

[85] 2006 Regulation on Administration of Water Abstraction Permits and Water Resources Fee Collection (China State Council), http://www.gov.cn/zwgk/2006-03/06/content_220023.htm (accessed May 21, 2017), art. 25.

[86] 2008 Measures on Administration of Water Abstraction Permits (China Ministry of Water Resources), http://www.gov.cn/flfg/2008-04/17/content_947055.htm (accessed May 21, 2017), art. 27.

and this system should ensure strengthening the security and certainty of water abstraction rights.

6.2.2 *Predictability and Priority*

The variability of water resources over time and space makes water availability difficult to predict. However, water rights should have a certain level of predictability so that right holders can have a reasonable expectation about the water volume that they can receive under their rights. Predictability of water volume can be affected by the way water rights are expressed. In California and Colorado in the US, water rights are expressed as a fixed volume of water, where a water right holder's likelihood of receiving the specified volume under their water right is subject to the time-based priority of right.[87] If water shortage takes place and there is no sufficient water supply available for all users, low priority right holders with later-dated rights have to bear the risks of not being able to satisfy all their water needs.

In other cases, instead of having the exact water volume specified, water rights are expressed as a share, such as a certain percentage of the total available resources. This approach allows the risks of unpredictability to be shared across all water rights holders.[88] In NSW, Australia, for example, access licences other than those of local and major urban suppliers hold shares and extraction rates determined by the rules of Water Sharing Plans.[89] In the event of water shortages, all right holders may receive a level of water supply based on their shares. In some state jurisdictions of Australia including Queensland, NSW, Victoria and Tasmania, a reliability system is in place for defining higher and lower reliability of water entitlements.[90] Reliability refers to "the frequency with which water allocated under a water access entitlement is able to be supplied in full".[91] Under different seasonal conditions, volumes of water allocation to high- and low-reliability entitlements can be different. Water is first allocated to high-reliability entitlements before any is allocated to lower reliability entitlements.[92]

Notwithstanding the difference in specifying water under a right, water rights arrangements in all jurisdictions have to define priority of water use for different types of uses to ration variability of water supply. In jurisdictions with the prior appropriation doctrine such as California and Colorado in the US, there are state

[87] Australia Productivity Commission, *Water Rights Arrangements in Australia and Overseas*, 99.

[88] Ibid.

[89] Ibid, 117.

[90] Australia National Water Commission (NWC), *Australia Water Markets Report 2012–13*, 2013, Canberra: National Water Commission, 19.

[91] NWC, *Water Markets in Australia,* 137.

[92] NWC, *Australia Water Markets Report 2012–13*, 19.

preference laws applicable to require that the domestic use and then irrigation use are satisfied first, even though the application rarely occurs in practice.[93]

Domestic use of water has generally been given higher priority over other use types, in first or second place after environmental use.[94] In NSW, for example, domestic and stock rights have second priority in water use after environmental water, except in severe water shortages.[95] In a circumstance that the Minister of Natural Resources declares a severe water shortage, domestic and basic rights then take higher priority over the environment during such a declaration.[96]

The priority given to the environmental water use is vastly different among jurisdictions. While some jurisdictions, including most Australian states, put it in the first place, the western US states treat the environment as equal with other use types for which the environmental rights to water also need to be subject to the time-based priority. In these jurisdictions, governmental authorities or environmental organisations can purchase senior water rights on the market and put the water to instream use for environmental protection.[97]

China's water abstraction rights are specified as a fixed volume of water which is subject to the supervision by authorities in charge of examining and approving the permits at annual intervals.[98] At the end of each year, abstractors have to submit an annual water abstraction report as well as advice on water abstraction plans for the subsequent year.[99] Authorities are responsible for preparing the water abstraction plans in a subsequent year for each abstractor based on a combination of the advice they submit, the annual water quantity allocation schemes, and the water abstraction plan for the administrative region.[100] The annual amount of water that can be abstracted under a permit might be restricted under any one of the circumstances as follows[101]:

- Water availability fails to satisfy the normal water supply for the region because of natural reasons.
- The act of water abstraction and return has severe negative impacts on the use of water areas, ecology, and the environment in water functioning zones.

---

[93] Australia Productivity Commission, *Water Rights Arrangements in Australia and Overseas*, 120. Also see David H. Getches, *Water Law in a Nutshell*, 3rd ed. (St. Paul, Minnesota: West Publishing Co., 1997).

[94] Australia Productivity Commission, *Water Rights Arrangements in Australia and Overseas*, 117–121.

[95] Ibid.

[96] Ibid, 249.

[97] Bates et al., *Searching out the Headwaters*, 157.

[98] 2006 Regulation on Administration of Water Abstraction Permits and Water Resources Fee Collection, art. 24, art. 40.

[99] Ibid, art. 42.

[100] Ibid, art. 39, art. 40.

[101] Ibid, art. 41.

- Groundwater is seriously over-extracted or ground extraction causes geographical disasters such as land subsidence.
- Other specific circumstances take place which need to limit water abstraction.

Furthermore, authorities have the power to restrict water abstraction in emergencies in severe droughts.[102] These provisions allow government to modify water abstraction rights in response to unpredictable climatic or other changes.

Different types of water use have been prioritised in China's water legislation. As described earlier by comparisons between the old and amended water law, environmental water use has obtained legal recognition with an increasing priority while domestic use as the first priority has been reaffirmed in both the 2002 Water Law and the 2006 Regulation on Administration of Water Abstraction Permits and Water Resources Fee Collection.[103] According to the regulation, local governments are mandated to set their local rules for water use to prioritise other uses except domestic use.[104] In Beijing, for instance, the environment and ecosystems have priority over other uses following domestic use, and industrial uses hold a higher priority over agriculture use.[105] By comparison, Tianjin government provides industrial use and agriculture use higher priorities (next to domestic use) over other uses.[106]

It is important to note that generally in practice, agricultural use and environmental use have for a long time been at lower priority than urban and industrial uses in China as economic development has been given the first priority. Urban and industrial uses make more contribution to the national GDP, and therefore they are favoured over agriculture by government. On the other hand, agriculture is also of strategic importance for China for food security in that its demands for water cannot be neglected. While agricultural water is taken away to satisfy the growing urban cities and booming industries, agriculture takes water that is supposed to be kept for the environment and ecosystems. In this sense, it is ultimately the environment that pays for economic growth.

As mentioned earlier, the total volume control of water abstraction is yet to be effectively implemented in China. For a long time, water abstractors have been less accountable for their acts of abstracting water beyond their entitled volume.[107] Similarly, when a water shortage takes place, water abstractors might not reduce the amount of water they abstract, for which the environment again takes the

[102] Ibid.

[103] For further information regarding the priority of water use, please refer to Sect. 5.3.3.3.

[104] 2006 Regulation on Administration of Water Abstraction Permits and Water Resources Fee Collection, art. 5.

[105] 2004 Measures on Implementation of the Water Law of People's Republic of China for Beijing Municipality (Standing Committee of Beijing Municipal People's Congress), http://www.bjcg.gov.cn/lawsearch.bbscs?action=read&idStr=119 (accessed September 28, 2008), art. 13.

[106] 2002 Regulation on Water Conservation for Tianjin Municipality (2005 Amendments) (Standing Committee of Tianjin Municipal People's Congress), http://www.chneef.com/law/qyfg/200504/20050423054233.html (accessed September 28, 2008), art. 17.

[107] Insufficient metering and monitoring infrastructure is another contributing factor to this situation.

consequences. Therefore, there are no clear priority rules in China to distribute the risk of unpredictability of water supply among abstractors, particularly those who are within a same water use category. This situation will hopefully change as water planning and water quantity allocation has made significant progress based on which the volume control of water use can be effectively implemented. To explore an appropriate solution to establish a primary rule of water rights priority, it is necessary to first take a look at the conditions of water abstraction permits.

After the examination of a permit application, the responsible authority issues a document of approval which includes[108]:

- the state of water quantity and quality of the water source, the purpose of water abstraction, the volume of water abstraction, and its corresponding guarantee rate;
- the location to which the return flows are disposed of, the volume of return flows, and the quality requirement of the return flows;
- the water use quota and the relevant water saving requirement;
- the requirement of water metering devices;
- measures of water abstraction restriction in particular circumstances;
- the requirement for water storage projects and hydropower projects regarding the water quantity distribution and reasonable flow quantity for lower reaches;
- matters of the application and other matters for attention.

The applicant who has been issued the approval document is able to commence the construction of the water abstraction works.[109] In the case that the construction project needs an administrative approval, the authority in charge of approving the project cannot accept the application without a water abstraction approval.[110] After the water abstraction works have been completed and passed the examination, the applicant is finally issued a water abstraction permit which includes following conditions[111]:

- The title or name of the abstractor
- The duration of the abstraction
- The volume of water for the abstraction and the purpose of use; the water volume refers to the maximum volume that can be abstracted under the condition of the average water availability over years
- Types of water resources
- The locations from which the water is abstracted and to which the return flows are disposed of, the return flows' pattern and quantity.

Among these conditions, the volume of water abstraction and its corresponding guarantee rate in the approval document can be viewed as a quantification of water

---

[108] 2008 Measures on Administration of Water Abstraction Permits, art. 21.

[109] 2006 Regulation on Administration of Water Abstraction Permits and Water Resources Fee Collection, art. 21.

[110] Ibid.

[111] Ibid, art. 24.

abstraction rights by which the former is a nominal volume and the latter represents the probability that the nominal volume will be received in full. However, the guarantee rate of water abstraction volume is not recognised by the permit itself, which only defines the maximum amount for water abstraction under the permit. In other words, this specified volume serves as a cap for the abstraction. Therefore, it can be found that permits do not explicitly specify the means by which permit holders within a same water use type share the risk of water shortage.

As an essential element of water rights, priority should be clearly defined either way (by prior appropriation or by a share). Some researchers call for the application of the prior appropriation doctrine in this context[112] It is proposed to introduce the time-based priority rule in China. Furthermore, water abstraction rights' priority should be based on the time of the permit applications rather than the time of the appropriation acts.[113] This suggestion is consistent with China's practice of permits as an origin of water rights. According to this argument, a right holder with a senior right would have higher priority to receive water supply in a water shortage than others who put water in the same type of use.[114] This proposal could be legitimately adopted by including a priority date in water abstraction permits. However, the time-based priority arrangement is not only related to distributing the risk of water shortage but has significant implications for water rights trading. For this reason, it should be carefully considered whether it is a potential solution for China in the context of promoting water markets.

Currently in practice, because a water abstraction right is established by issuing a permit, a new water user is unable to keep the priority date of the water abstraction right that they have purchased, even though this right is originally a senior right. The priority of this right changes, or actually loses its seniority, after transfer. Put another way, the new right holder can only obtain a right with its priority based on the time of their new permit. This current practice is obviously in conflict with the purpose of the time-based priority arrangement.

A time-based priority differentiates the value of water rights. Due to the difference in the probability of receiving water in a shortage, a senior water right should theoretically have higher value in the market than a junior water right. In China, however, the new water user who purchases a water right, regardless of the original priority of the right purchased, acquires a new permit, which means that the priority of this water abstraction right becomes low in terms of the time-based priority proposal. From this perspective, a senior right and a junior right would have no difference in values for the purchaser, which again is inconsistent with the time-based priority arrangement.

Water abstraction permits are subject to renewal at the end of duration. When an abstraction right is renewed, the right holder will be issued a new permit which includes a new duration.[115] By this means, the right holder loses their priority date

---

[112] Cui, *Studies on Quasi Property Rights*, 318.

[113] Ibid, 308.

[114] Ibid, 316.

[115] 2008 Measures on Administration of Water Abstraction Permits, art. 27.

established by the previous permit. In other words, a right holder whose existing permit has been renewed would be in the same situation as a new water user whose application for a permit has been approved. This practice apparently denies the rule of time-based priority.

If the proposal is to be adopted in China, all these conflicting practices will have to be addressed. However, it seems practically difficult as it involves a fundamental modification of the current legal and institutional framework, which is less likely to occur. Based on the above analysis, therefore, it is concluded that it is unfeasible to arrange the time-based priority rule under the current framework of the use type-based priority.

6.2.3 Government Intervention in the Exercise of Water Rights

Government's power to interfere with the exercise of water rights varies across jurisdictions. Generally speaking, governmental authorities in the western US and Australia do not have the power to cancel or suspend a water right without compensation during its term for reasons other than enforcement.[116] In Australia, the federal government has made significant financial commitments ($3.1 billion over 10 years) to buy back permanent water entitlements from farmers to recover water for the environment.[117] Such a buyback programme enables farmers to be fully compensated for selling their water entitlements.[118]

Government in these jurisdictions has various responsibilities to administer water rights. In NSW, for instance, water rights can be compulsorily taken by the government based on the rule of public interests.[119] In addition, where there is rapid population growth, the Minister for Natural Resources of NSW may modify the water right of a major or local water infrastructure service provider by varying it at five-yearly intervals.[120] If the modification causes a reduction in water availability to a licensed water user during a term of plan which authorises the water right, the user is entitled to compensation.[121] However, no compensation is payable for the reduction as a result of a new plan.[122]

In China's current legal system, there are no clear provisions to regulate government's interference with existing abstraction rights. Instead, the 2007 Property Law

---

[116] Australia Productivity Commission, *Water Rights Arrangements in Australia and Overseas*, 106.

[117] NWC, *Water Markets in Australia*, 78.

[118] NWC, *The National Water Initiative – Securing Australia's Water Future: 2011 Assessment*, 2011, Canberra: National Water Commission, 2011, 127.

[119] Australia Productivity Commission, *Water Rights Arrangements in Australia and Overseas*, 106.

[120] Ibid, 207.

[121] Ibid.

[122] Ibid.

explicitly entitles the usufruct right holders with the right to claim compensation for a governmental expropriation or requisition. As Article 121 in the Property Law states:

> Where the usufructuary right is terminated or its exercise is affected because of expropriation or requisition, the usufructuary right holder has the right to obtain corresponding compensations is accordance with Articles 42 and 44 of the present law.[123]

Based on Articles 42 and 44, compensation is payable for lands, other real estates and chattels expropriated or requisitioned for the needs of public interests or emergencies.[124]

The Property Law recognises water abstraction rights as a type of usufructuary right, and thus it can be interpreted that if water availability under a water abstraction right is reduced or taken by government for a public interest during its existence, right holders should be entitled to corresponding compensation. An explicit addition of such provision in the Water Law would help increase the security of water abstraction.

Government interference with the exercise of water rights is also reflected in conditions of water abstraction permits. Abstractors are obliged to reapply for a new permit if they would like to make critical changes to the permit conditions during the existence of the permit. The following changes are considered critical[125]:

- Changes in the water volume for abstraction, or changes in the purpose for which the water is put for; however, in the case that the changed volume is a result of water abstraction right transfer, it does not need a reapplication
- Changes in the water source or the location from which the water has been abstracted
- Changes in the location to which return flows are disposed of, the quantity of return flows, or the patterns of return flows

[123] 2007 Property Law of the People's Republic of China, http://www.npc.gov.cn/englishnpc/Law/2009-02/20/content_1471118.htm (accessed May 21, 2017), art. 121.

[124] Article 42 in China 2007 Property Law states that, "For the needs of public interests, collective-owned lands, premises or other real estates owned by entities and individuals may be requisitioned in accordance with the statutory power and procedures. When requisitioning collective owned lands, the full amount of land compensation fees, relocation subsidies, compensations for the above-ground fixtures of the lands and seedlings, and other fees are payable in accordance with law. Social security fees must be arranged for the farmers with the land requisitioned, and their livelihood and lawful rights and interests must be guaranteed and protected. When requisitioning premises or other real estates owned by entities and individuals, it is required to compensate for demolishment and relocation in accordance with law and protect the lawful rights and interests of the owners of the requisitioned realties. When requisitioning the individual residential homes, it is required to guarantee the housing conditions of the owners of the requisitioned homes…" Article 44 states that, "For the needs of emergent dangers or disasters, it is allowed for one to use the realties or chattels owned by entities and individuals according to the statutory power and procedures. Such realties or chattels shall, after the emergent use, be returned to the owners. Corresponding compensation shall be payable for realties or chattels being used, damaged or lost because of the use".

[125] 2008 Measures on Administration of Water Abstraction Permits, art. 29.

- Changes in the major pollutants in return flows and measures of wastewater treatment.

This provision can be viewed as use restrictions of water abstraction permits, which means that rights can only be exercised for specific purposes and in a certain location. It allows government to keep a strong control over water use and allocation, but at the same time gives right holders little flexibility.

6.2.4 Transferability and Divisibility of Water Rights

Making water rights tradable or transferable can serve as a management tool to achieve efficiency as well as flexibility of water allocation. The creation of a price signal encourages efficiency in water use, for which tradable water rights allow water users to find the highest value for their water.[126] Flexibility, on the other hand, has been increasingly taken into consideration in water allocation to manage the variability of water resources and the risks of drought. In response, trading in water rights, particularly short-term or temporary water rights trading, has been recognised as an important tool for flexible water management.[127]

According to water trading experiences across overseas jurisdictions, most water rights trades are temporary rather than permanent, such as trades in annual water assignments and short-term leases.[128] In Australia, for instance, 28,899 temporary allocation trades were recorded nationally, compared with 9087 permanent entitlement trades in 2014–15.[129] Among permanent entitlements traded in Australia, environmental purchases by the Australian Government accounted for a considerable proportion.[130] In California, the success of the Drought Water Bank in 1991 and 1992 was to a large extent attributed to the preference of the agricultural community to short-term transfers.[131] Even in the Chilean water market where water rights as a private real property are given extremely high level of legal security, trades more frequently took place on a short-term basis within the agricultural sector as a tool for farmers to manage risks associated with uncertainty in water supplies and in agricultural markets.[132]

[126] Debra Townsend and Odette Adams, "Water Rights and Trading in Australia," *Australian Environmental Law Digest*, September (39): 3.

[127] David J. Guy, "A Model Water Transfer Act for California: An Agricultural Perspective," *Hastings West-Northwest Journal of Environmental Law & Policy* 14 (40): 722.

[128] Australia Productivity Commission, *Water Rights Arrangements in Australia and Overseas*, 136.

[129] Australia Bureau of Agricultural and Resource Economics and Sciences (ABARES), *Australian Water Markets Report 2014–15*, 2016, Canberra: Department of Agriculture and Water Resources, 18.

[130] NWC, *Australian Water Markets Report 2012–13*, 9.

[131] Guy, "A Model Water Transfer Act for California," 714–715.

[132] Carl J. Bauer, *Siren Song: Chilean Water Law as a Model for International Reform* (Washington & Great Britain: Resources for the Future, 2004), 89.

Accordingly, it has been observed that water law and policy reform attempts to facilitate the flexibility function of water rights trading. In Australia, state jurisdictions agreed in the NWI to establish water entitlements consistent with the NWI definitions. While terminology and stage of implementation vary across Australia, a set of national definitions for "water access entitlement" and "water allocation" are provided under the NWI for reform in water rights arrangements. Water access entitlement refers to "a perpetual or ongoing entitlement to exclusive access to a share of water from a specified consumptive pool as defined in the relevant water plan".[133] Water allocation refers to "the volume of water allocated to a water access entitlement in a given season, defined according to rules established in the relevant water plan".[134]

The relationship between water access entitlements and water allocations is as such:

> In a number of jurisdictions, water allocations are bundled with water access entitlements in Australia such that the water access entitlement gives the holder a right to access water and the water allocation specifies how much water the holder can access.[135]

Water access entitlements in Australia are established as tradable through the separation from land ownership. Components in the water access entitlement are identified including a long-term interest in a (share of) a consumptive pool as defined in a water plan, a unit of opportunity (usually a volume of water) distributed periodically, the delivery right, and use (water use licence).[136] These traditionally bundled elements of a water right are further unbundled as a next step of untying water rights from land rights.[137] Through this process of unbundling, water rights are made divisible, potentially separately tradable, to maximise the benefits of trade.[138] The following table (Table 6.2) summarises the status of unbundling of water rights across Australian state jurisdictions.[139] Unbundled water rights, depending on their tradability, may be transferred permanently or temporarily, fully or partially. The transfer can be also absolute or for a defined period under a lease arrangement.[140]

A variety of water trades has also been promoted in the western US by providing strong protection to water rights. Conservation-based transfers in California can be provided as a noteworthy example. The rule of beneficial use requires water to be

---

[133] NWC, *Water Markets in Australia*, 138.

[134] Ibid.

[135] Australia Bureau of Statistics, *2004–05 Water Access Entitlements, Allocations and Trading*, http://www.abs.gov.au/ausstats/abs@.nsf/mf/4610.0.55.003 (accessed September 28, 2008), 7.

[136] NWC, *Water Markets in Australia*, 83.

[137] Ibid.

[138] Australia Productivity Commission, *Water Rights Arrangements Australia and Overseas*, 112.

[139] NWC, *The National Water Initiative – Securing Australia's Water Future*, 29. Also see NWC, *Australian Water Markets Report 2012–13*.

[140] NWC, *Australian Water Markets Report 2012–13*.

Table 6.2 Unbundling water rights across Australian jurisdictions

Jurisdiction	Entitlement	Tradability
NSW	Unbundled:	
	Water access licence	Yes
	Water allocation	Yes
	Water supply works and use approval	No
	Bundled (unregulated surface water and groundwater): Water Licence under the Water Act 1912	Yes if provided for by the relevant water sharing plan
Victoria	Unbundled (regulated river water):	
	Water share	Yes
	Water allocation	Yes
	Delivery share	Yes
	Water-use licence	No
	Bundled (groundwater and unregulated surface water): Groundwater and unregulated surface water licences under the Water Act 1989	Yes subject to trading rules
Queensland	Unbundled:	
	Water allocation	Yes
	Seasonal water assignment	Yes
	Bundled:	
	Interim water allocation under the Water Act 2000	No
	Water licence (unregulated surface water and groundwater)	No
South Australia	Unbundled:	
	Water access entitlement	Yes
	Water allocation	Yes
	Water resource works approval	No
	Site use approval	No
	Bundled:	
	Water licence under the Natural Resource Management Act 2004	Yes
Western Australia	Water licence under the Rights in Water and Irrigation Act 1914 (unbundled from land but otherwise bundled)	Yes
	Water allocation (only applicable to cooperatives)	Yes
Northern Territory	Water licence under the Water Act 1992 (unbundled from land but otherwise bundled)	Yes in areas covered by water allocation plans
Tasmania	Unbundled from land:	
	Water licence under the Water Management Act 1999	Yes
	Water allocation	Yes
	Irrigation right	Yes within the irrigation district

(continued)

Table 6.2 (continued)

Jurisdiction	Entitlement	Tradability
Australian Capital Territory	Unbundled from land under the Water Resources Act 2007:	
	Water access entitlement (a right to take a specified volume of surface water or groundwater)	Yes
	Water licence (specifies the location from which water can be taken)	No. A water licence holder must hold water access entitlement(s).
	Water allocation	Yes

(Data from NWC 2011b, the National Water Initiative – Securing Australia's Water Future; NWC 2013, Australian Water Markets Report 2012–2013)

continuously used otherwise water rights may be lost in accordance with the forfeiture doctrine, which discourages water conservation.[141] The California Water Code section 1011 states that an appropriate water right holder "who fails to use a portion of that water, due to conservation measures, will be deemed to have put that conserved water to beneficial use. The conserved water, or the right to its use, may then be transferred to another water user".[142]

Furthermore, short-term transfers have been promoted by making water rights holders more secure in their water rights and the future ability to use the water.[143] For instance, the Model Water Transfer Act for California provides that "the transferor's water right will not be diminished during the terms of the transfer and at the conclusion of the transfer agreement, the water reverts back to the transferor".[144] By comparing this provision to Australian unbundling of water rights, it can be observed that it is a facilitation of temporary trades, similar to water allocation trades in Australia.

Temporary trades in water rights in China are not yet well facilitated by the current systems of water abstraction permits. According to Article 27 of the Regulation on Administration of Water Abstraction Permits and Water Resources Fee Collection, China only allows conservation-based water rights trades. The current trading practices in China mainly represent new water users who want long-term water abstraction rights. In the Yellow River Basin, while a water abstraction permit could have a duration shorter than 5 years,[145] the duration of a transfer agreement can be as long

[141] Getches, *Water Law in a Nutshell*, 176–179.

[142] Scott S. Slater, "A Prescription for Fulfilling the Promise of a Robust Water Market," Symposium on the 25th Anniversary of the Report of the Governor's Commission to Review California Water Rights Law, *McGeorge Law Review* 36 (2005): 261.

[143] Guy, "A Model Water Transfer Act for California," 716.

[144] The Model Act section 101, quoted from Ibid.

[145] The duration of no more than 5 years was according to Article 26 of the 1994 Implementing Rules on Water Abstraction Permits in the Yellow River Basin (YRCC). After the 2006 Regulation on Administration of Water Abstraction Permits and Water Resources Fee Collection was released, however, the old permits with less than 5 year duration were required to convert into new ones with durations between 5 and 10 years in accordance with Article 25 of the new Regulation.

as 25 years. Such maximum duration for a water right transfer is based on Article 18 of the Implementing Rules on Administration of Water Rights Transfer in the Yellow River Basin (Trial) released by the YRCC in 2004:

> The duration of water rights transfers shall be determined with balancing interests of both parties and based on overall consideration to the life expectancy of the major water saving project, the term for renovation of the transferee's major project, and changes in the water market and water resources allocation of the Yellow River Basin. In principle, the duration for a water right transfer shall not exceed 25 years...When the duration expires, the transferee who needs to continue water abstraction shall arrange a new water right transfer. In the case that the transferee does no longer abstract water, the water right reverts back to the transferor, and the amount of water under the right is subject to the readjustment by authorities in charge of water abstraction permit examination and approval.[146]

As already mentioned, water rights transfers in the Yellow River Basin usually took place between irrigation districts and power plants.[147] New power plant developers invest in water conservation rehabilitation projects in irrigation districts and acquire water allocation quotas saved from irrigation through these investments. In these water rights transfer agreements, transferors were irrigation district authorities who often are agencies of water administration departments. In charge of issuing water abstraction rights in its jurisdiction, the YRCC took the initiative to promote the pilot transfers to satisfy new industrial water demands while keeping water use within the quota limit. Therefore, water abstraction rights for these transfers had higher security than others in terms of little risk of attenuation or loss at the end of permit duration. Put in another way, the transferee can be reasonably confident that the water abstraction right obtained from the transfer will be successfully renewed when the permit duration expires within the 25 years.

The 2016 Provisional Measures on Administration of Water Rights Trading similarly provides that a transfer of water abstraction right can have a duration longer than that of the permit, as

> The water authority in charge of examining and approving the transferee's new water abstraction permit application shall, in consultation with the authority that issued the original permit, check and ratify the duration of the transfer, and specify it in the approval document. Within the ratified duration, the transferee's water abstraction permit shall be extended with a high priority, except that the transferee failed to submit an application for extension in accordance with law.[148]

Such provisions reflect on China's water trading practices that water abstraction right transfers tend to be long term if not permanent. If a water abstractor transfers all or part of their right, the amount of water that can be taken or received under the permit will be changed according to the transfer. The transferee, on the other hand,

---

[146] China Ministry of Water Resources, Water Management Department, ed. *The Report on Pilot Experiences of Establishing the Water Rights System Part 1: Materials of Water Rights Transfer in the Inner Mongolia Autonomous Region* (Beijing: China Water Publishing Co., 2005), 102.

[147] For details about the water rights transfers in the Yellow River Basin, refer to Sect. 4.4.3.

[148] 2016 Provisional Measures on Administration of Water Rights Trading (China Ministry of Water Resources), http://www.gov.cn/zhengce/2016-05/22/content_5075679.htm (accessed May 27, 2017), art. 16.

would gain the right by applying for a corresponding permit.[149] Although a permit can only last for 5–10 years, the transferee can apply for permit extension within the duration of the transfer, with a high level of confidence that their application will be most likely successful. The 2016 Provisional Measures on Administration of Water Rights Trading does not set an upper limit for the duration of water abstraction rights transfer, which implies that a water abstraction right transfer could be longer than the previously set Yellow River Basin limit of 25 years. While this approach allows limited water resources to become available for new water users and encourages efficiency in water allocation, it fails to provide flexibility. In order to augment the benefits of water rights trading, there is space for China to learn from overseas experiences in the implementation of temporary trades. Annual water abstraction plans[150] within the current system can potentially serve as a base to build up the rules of temporary water trades. In other words, annual water abstraction plans may be unbundled from water abstraction rights so that all or part of the annual amount of water under the plan becomes transferrable between abstractors. The transfers can be one year long or shorter (seasonal) to satisfy water demands of different abstractors affected by temporal water variability. By only transferring water allocation under the annual water abstraction plan, water rights holders are able to keep their water abstraction rights (permits) for future use.

6.3 Defining Tradable Water Rights for the South-to-North Water Diversion Project

As noted in Chap. 4, the theory of water rights has been introduced in the construction and administration of the SNWDP although it originally came up as a technological and administrative allocation option. Proposed to be run with a mixed management approach, the SNWDP is expected to become the largest water rights trading market around the world. How to design tradable water rights for this project is of great importance to the application of the market allocation mechanism in China. Different from natural water systems, water rights arrangements for the SNWDP are such a complicated legal issue that is yet to be fully addressed in the literature. Previous studies have recognised that a water supply corporation should be established in which provincial governments make their investments according to their water needs.[151] Rather than holding their own water abstraction rights, provincial governments hold the corporation shares upon which their water use quotas are determined.[152] Although these research outcomes have innovatively attempted to clarify the water rights arrangements between the state and the project, as well as

[149] The right, however, is still subject to the provisions pertaining to the duration of water rights, government review, and other governmental interference.

[150] For more details about the annual water abstraction plans, refer to Sect. 5.3.2.2.

[151] Cui, *Studies on Quasi Property rights*, 355.

[152] Ibid, 356.

among provincial governments involved in the project,[153] the existing body of literature lacks a comprehensive exploration on the tradability of water rights involved in the project. This critical knowledge gap needs to be addressed in order for the water rights trading market to grow in the context of the project.

6.3.1 The Ministry of Water Resources Proposal

6.3.1.1 Geographical Perspective

The Eastern Route and the Middle Route of the SNWDP have been in service since 2013 and 2014. Starting from Danjiangkou Reservoir in the middle reaches of the Yangtze River, the 1267 km long canal of the Middle Route goes through Henan Province and Hebei Province, and reaches Tuancheng Lake in Beijing. To divert the southern water to another thirsty northern city, Tianjin, a 154 km long canal connects the main Middle Route canal to the Waihuan River in Tianjin.[154] Phase I of the Middle Route was proposed to annually divert an average amount of 9.5 billion cubic metres of water from the Yangtze River to four water-receiving provinces and municipalities, namely, Henan Province, Hebei Province, Tianjin, and Beijing.[155]

Diverting water from the main course of the Yangtze River in Jiangsu Province, the Eastern Route goes north along the ancient Grand Canal and watercourses in parallel with it. As shown in the following map, it connects four lakes on the way, namely, Hongze Lake, Luoma Lake, and Nanci Lake, and Dongping Lake, and divides into two lines from Dongping Lake. One line crosses the Yellow River, goes north through Hebei Province, and reaches Beidagang Reservoir in Tianjin, and the other line goes east to Mishan Reservoir in Weihai City, Shandong Province.[156] Focusing on the sections in Jiangsu and Shandong, the first phase of the Eastern Route was proposed to supply 3.9 billion cubic metres of Yangtze water per year to the two receiving provinces.[157]

The specific local context of the two routes has led to different institutional arrangements for administration, which have impacts on shaping the water rights arrangements. To explore the design of tradable water rights for the SNWDP, it is helpful to analyse these institutional frameworks.

---

[153] Ibid, 355–360.

[154] Data from the official website of the Henan Section of the South-to-North Project, http://www.hnnsbd.gov.cn/index/gaikuang.asp (accessed September 28, 2008).

[155] Data from the official website of the South-to-North Water Diversion Project, http://www.nsbd.gov.cn/ (accessed September 28, 2008).

[156] Ibid.

[157] Ibid.

6.3.1.2 The Middle Route

The MWR developed a proposal which suggested the establishment of a Main Line Corporation for the South-to-North Middle Route was proposed as a joint-ownership enterprise where the state holds a controlling share and the four areas concerned in the route hold their corresponding shares based on their investments.[158] The Corporation as a legal person, is responsible for managing the construction and operation of the main line for the project.[159] In 2004, the central government invested RMB 300 million yuan to establish the Construction and Administration Bureau of the South-to-North Water Diversion Middle Route Project to work as the original basis for the corporation (hereinafter the Middle Route Bureau).[160]

In addition to the Middle Route Bureau, the Water Source Corporation of the South-to-North Middle Route (hereinafter the Middle Route Source Corporation) was set up in the same year to take the responsibilities of constructing and managing projects related to the water source of the route, such as the heightening of Danjiangkou Dam, the construction of the Taocha Diversion Hub, and the relocation of the people in the reservoir area.[161]

The Middle Route Bureau and the Middle Route Source Corporation create an additional administrative level between the state and provincial governments. Because of these artificial institutions different from the usual administrative hierarchy, the "three step" water rights initialisation framework analysed earlier in Sect. 5.3 found itself less suitable for the MRS context. This institutional arrangement raised two concerns: how is the water usufruct right separated from the state ownership? Which organisation, the Bureau or the Corporation, should be the holder of the water abstraction right? Answers to these two questions determine three relationships, namely between the state and the Bureau, between the state and the Source Corporation, and between the Bureau and the Source Corporation.

According to the MWR proposal, the relationship between the Bureau and the Source Corporation should be based on water supply contracts.[162] This proposal was developed 3 years earlier than the 2006 Regulation on Administration of Water Abstraction Permits and Water Resources Fee Collection when water abstraction permits were recognised as legal basis for water rights. As a result, water rights arrangements were not addressed in the proposal. The recommendations made in the proposal did attempt to incorporate market mechanisms into the traditional

---

[158] China Ministry of Water Resources, Development Research Centre, "An Introduction to Research on the Construction and Administration System of the South-to-North Project," *China Water Resources* 1B, (2003): 72.

[159] Data from the official website of the Construction and Administration Bureau of South-to-North Water Diversion Middle Route Project, http://www.nsbd.cn/n524903/index.html (accessed August 27, 2008).

[160] Ibid.

[161] Data from the official website of the Water Source Corporation of the South-to-North Middle Route, http://www.zxsygs.com/html/gsgk/ (accessed August 27, 2008).

[162] Development Research Centre, "An Introduction to Research on the Construction and Administration System of the South-to-North Project," 73.

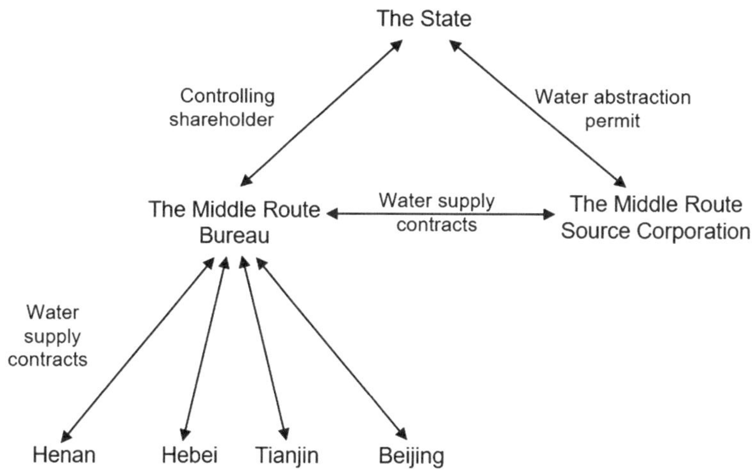

Fig. 6.1 MWR proposed water rights arrangements for the South-to-North Middle Route

framework of the administrative approach to water allocation, but failed to link the institutional arrangement for the Middle Route to the water abstraction permits system.

Based on this proposal, the three relationships were designed as such (shown in Fig. 6.1):

- The Middle Route Source Corporation holds a water abstraction permit based on which the usufruct right to water can be established from the state's water ownership.
- The Middle Route Bureau or the future Main Line Corporation receives water from the Source Corporation based on a water supply contract. Provincial governments hold shares in the Bureau and corresponding water use quotas to the project water. These quotas represent a share of the water supply contract rather than that of a water abstraction permit.
- As the controlling shareholder of the Bureau or the future Main Line Corporation, the state does not claim any volume of the project water for abstraction but takes control over it for the sake of public interests.[163]

The Source Corporation stands before the Bureau in terms of its responsibilities for managing the construction and operation of Danjiangkou Reservoir which is the starting point of the project. It might be a more effective arrangement to set up two separate organisations to respectively construct and operate the water source project and the main line project in terms of engineering management. The water source project is based on the existing Danjiangkou Reservoir, which historically fell in the jurisdiction of the Han River Group of Water Conservancy and Hydroelectric Power (hereinafter the Han River Group). The Han River Group is a state-owned public

[163] Ibid, 72.

company, which is involved in the construction, operation, and management of the reservoir and its related water hub projects since 1958.[164] The new Source Corporation was then established based on the Han River Group. The capital investment from the central government for the Middle Route water source project directly goes to the Source Corporation. Furthermore, the added capital of the Source Corporation was not incorporated into the future Middle Route Corporation.[165] This arrangement implied that the proposal did not intend to make the Source Corporation the water right holder but the "project manager" for convenience.

However, the neglect of water abstraction permits (rights) in the proposal directed the design of the institutional arrangement away from its original intention. As a joint-ownership company, the Bureau was supposed to be the water right holder in which shareholders (except the state itself) can obtain entitlements to the water. These entitlements were intended to be shares to the usufruct right instead of a water supply contract. However, the initial intention was lost by separating functions of the prospective corporation into two organisations. Compared to the Source Corporation, the Bureau is in a better position to act as the water right holder. By a step-by-step approach, it can be converted into a jointly-owned corporation in which provincial governments including Henan, Hebei, Tianjin, and Beijing hold their shares and corresponding water entitlements. The Bureau as the water right holder seems to make more sense. To summarise, the proposed water supply contract arrangement failed to accommodate the original intention of the proposal.

In 2014, the State Council released the Regulation on Administration of Water Supply and Use for the SNWDP. The regulation provides that water supply contracts shall be signed between authorised governmental departments in water receiving provinces and municipalities and management units for the SNWDP determined by the State Council (Article 6 and Article 14).[166] In charge of the operation and maintenance of the SNWDP, these management units are also water abstraction permit holders. The new regulation therefore reaffirmed the water supply contract arrangement, which provides lower level of security and legal protection of water users than property right based entitlements. To achieve higher security for tradable water rights, an alternative proposal based on water abstraction rights is made here for the MRS:

- The Middle Route Source Corporation could either join the Han River Group as a subsidiary or merge into the future Middle Route Corporation, keeping its responsibility in the operation and maintenance of the water source related projects. Some scholars supported the second option from a perspective of the

[164] Data from the official website of the Han River Group of Water Conservancy and Hydroelectric Power, http://www.hjgrp.com/about_us/lsyg_yg.asp (accessed August 28, 2008).

[165] Development Research Centre, "An Introduction to Research on the Construction and Administration System of the South-to-North Project," 73.

[166] 2014 Regulation on Administration of Water Supply and Use for the South-to-North Diversion Project (China State Council), http://en.pkulaw.cn/display.aspx?cgid=219668&lib=law (accessed May 21, 2017), art. 14, art. 6.

integrated water management.[167] This option is arguable because it is politically difficult and inconsistent with the overall institutional framework for the Middle Route. As noted earlier, the investment for the water source project went directly to the Han River Group and its capital return stayed out of the assets of the Middle Route Corporation, which implied that the separation of the two organisations was a compromise on the existing state of interest distribution. In this sense, it seems that the Han River Group keeping the Source Corporation may be a more suitable solution because it would maintain the status quo and be more politically acceptable.
- The Middle Route Bureau or the future Main Line Corporation holds the usufruct water right in the form of the water abstraction permit from the state ownership.
- Both the state and provincial governments concerned in the Middle Route project jointly hold the shares in the Bureau or the future Main Line Corporation, which represent ownership of its infrastructure and assets. As the controlling shareholder, the state retains the power to the project and the water for protection of public interests.
- As shareholders in the Bureau or the future Main Line Corporation, provincial governments are served by the project based on their water entitlements in the form of the shares in the overall water abstraction permit.

Establishing property right based water entitlements at the provincial level in the context of the SNWDP may provide better facilitation for potential water rights trading activities between the water receiving provinces and municipalities. While regional water rights can be traded between county and above governments according to the 2016 Provisional Measures on Administration of Water Rights Trading, such regional water rights trading is limited within water allocation quotas for natural water systems (Article 3). However, the trades of SNWDP water supply contracts started occurring between local governments in Henan Province through the newly established China Water Exchange in late 2016. This is another example of emerging practices not consistent with legal provisions. To further grow water rights trading markets, there is clear need for reform in water rights arrangements in the context of the SNWDP.

6.3.1.3 The Eastern Route

The Eastern Route involves four provinces and municipalities, namely Jiangsu Province, Shandong Province, Hebei Province, and Tianjin. Jiangsu has its own diversion project to transfer the Yangtze River water to its Northern area for more than 40 years. The route section falling in Jiangsu's territory was extended based on

[167] Liwei Zhang et al., "A Review of the Australian Experiences in the Water Quality Security and Administration of the Snowy Mountains Project," *South-to-North and Water Conservancy Technology* 2, 2007, http://www.hwcc.com.cn/newsdisplay/newsdisplay.asp?Id=178716 (accessed September 28, 2008).

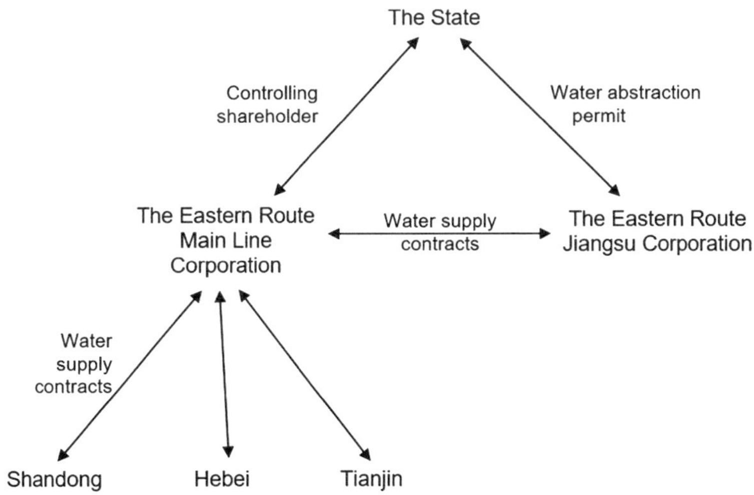

Fig. 6.2 MWR proposed water rights arrangements for the South-to-North Eastern Route

its existing diversion project. Similar to the Middle Route, the institutional arrangements for the Eastern Route also need to take the status quo into consideration. As the MWR proposal suggested, it was difficult to clearly define property rights between the previous projects and the extension ones. In addition, the existing projects serve a range of functions from water transfer, navigation, flood control, irrigation, to wastewater discharge, which add difficulties in operation and administration.[168] Accordingly, a separate approach to organisation became preferable over a single organisation approach. Two alternative arrangements were introduced and compared in the MWR proposal:

- To establish two Eastern Route water transfer corporations, one based on Jiangsu Province and the other on the other three provincial governments. The Eastern Route Jiangsu Section Corporation is responsible for the construction and administration of the extension projects within the territory of Jiangsu Province. Other sections of the route project remain in the jurisdiction of the Eastern Route Main Line Corporation shared between the state (with a controlling holding) and provincial governments concerned in the route, including Shandong, Hebei, and Tianjin.[169]
- To establish Eastern Route water transfer corporations based on each province or municipality concerned in the route, responsible for the construction and administration of each own route section. In addition to these provincial corporations,

[168] Development Research Centre, "An Introduction to Research on the Construction and Administration System of the South-to-North Project," 73.

[169] Ibid.

an Eastern Route Council at the central level consists of representatives from the MWR and the provincial governments to take a coordinative role.[170]

The proposal recommended the first approach. The project was constructed on a phase-by-phase basis, and the first phase only involved Jiangsu Province and Shandong Province. The Eastern Route Shandong Main Line Corporation was first established in 2004 for the construction and administration of the Eastern Route Phase 1.[171] The Jiangsu Water Source Corporation was also set up in 2005, in charge of operation of the project within the territory of Jiangsu Province. The above figure (Fig. 6.2) shows how the players in the Eastern Route are related to each other and where institutional arrangements are needed to create the relationships between the players.

Similar to the Middle Route, the contractual arrangement was made for the Eastern Route. As the MWR proposal suggested, between the Jiangsu Corporation and the Eastern Route Main Line Corporation (first as the Shandong Corporation) is a relationship based on water supply contracts.[172] Rather than having a water usufruct right, the Main Line Corporation (firstly as the Shandong Corporation for Phase I) can only claim water supply from the Jiangsu Corporation based on a contractual right. Water entitlements of other provincial governments represent also in the form of water supply contracts.

As already elaborated in the previous section on the Middle Route, the arrangement of two separate corporations may be justified from an engineering management point of view, yet it is more a compromised result to maintain the historic interest distribution. The separation approach intended to maintain the status quo by allowing Jiangsu Province to keep its property right to the existing water transfer projects. However, it went astray when designing the water rights arrangement. The water supply contract arrangement can be challenged for two reasons:

- Jiangsu Province is one of the investors and users of the Eastern Route project and should not be given a different position that may be more superior to the other three provinces and municipalities concerned in the project. The four provincial governments should be treated equally in the water rights arrangement. Failing to do so, the contractual arrangement gave Jiangsu the water usufruct right while the other three governments can only have less secure water entitlements based on water supply contracts.
- As explained in the context of the Middle Route, a jointly-owned corporation should be the one who holds the water abstraction right on behalf of its shareholders. This was the original intention for the institutional arrangements for the SNWDP, but was lost in the separation of multiple organisations. Water rights

---

170 Ibid.

171 Data from the official website of the South-to-North Water Diversion Project, http://www.nsbd.gov.cn/zw/zqxx/jsgl/3/20061018/200501070025.htm (accessed August 28, 2008).

172 Development Research Centre, "An Introduction to Research on the Construction and Administration System of the South-to-North Project," 73.

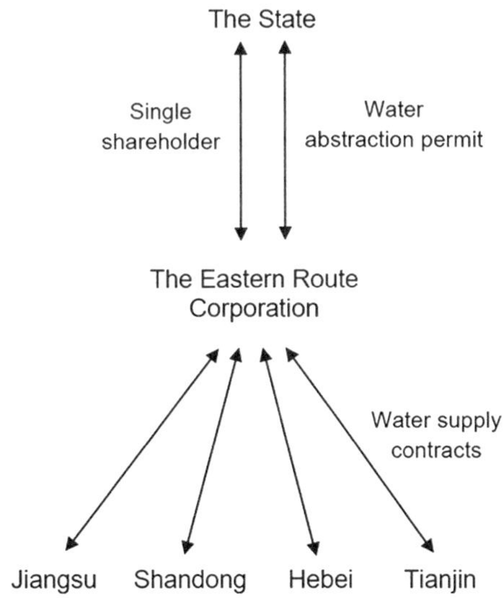

Fig. 6.3 Current water rights arrangements for the South-to-North Eastern Route

arrangements may have to respect the reality of historic interest distribution, but in the meantime, need to seek a balance among interested parties.

Based on the above analysis, therefore, it seems that there is a need to modify the design of the water rights arrangements for the Eastern Route. In 2014, the new Eastern Route Corporation was established to take over the role in operation and management of the ERS main project. With the RMB 10-billion-yuan investment by the State Council, the Eastern Route Corporation is a state-owned enterprise. The Jiangsu Corporation is no longer responsible for operation and management of the project, but for water supply of the project within the province. The modified water rights arrangement for the Eastern Route is shown in Fig. 6.3. It can be observed that the new water rights arrangement for the Eastern Route is a major modification which has moved away from the originally proposed separation approach. Instead of a water supply contractor, the Eastern Route Corporation has become a water right holder.

6.3.1.4 The Proposed Institutional Framework

The MWR proposal established an institutional framework that includes six levels of water entitlements for the SNWDP in which water supply contracts were employed as a major water right arrangement. Besides the main line projects diverting water from the Yangtze River and transferring it to the arid north of China, provinces and municipalities need to construct their accessory projects for the diversion and delivery of water to their users. Water supply or water distribution companies were established at the provincial level responsible for constructing, operating and managing the accessory projects within their territory, and supplying water to

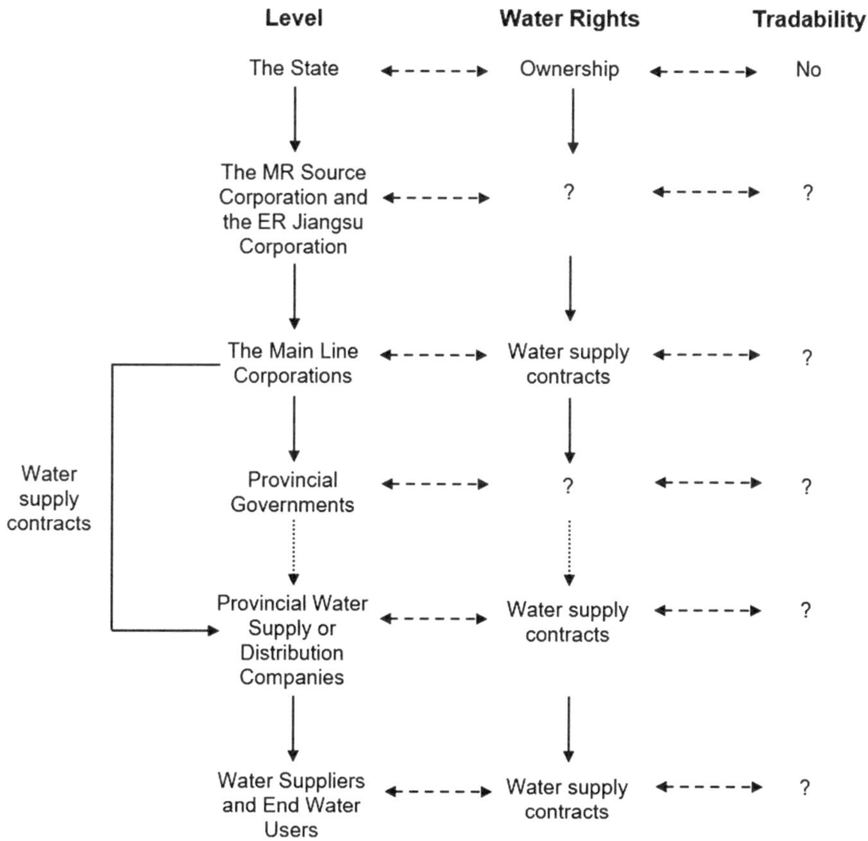

Fig. 6.4 Six levels of water rights in the South-to-North Water Diversion Project in China

their water users. According to the MWR proposal, the provincial water supply or distribution companies should be served by the Main Line Corporations (of the Middle Route and the Eastern Route) as water supply contractors.[173] The six levels of water rights are illustrated in Fig. 6.4.

It is shown that in this six-level framework, the legal status of water rights at the source corporation level as well as at the provincial level is less clearly specified. Tradability of the water rights across the levels remains similarly unclear. The six levels of organisations in the current design include from the state level down to the water user level, among which two separate corporations at the project level were set up as the initial arrangement. Such arrangement failed to recognise the usufruct right by creating an additional level of organisation (as Source Corporations) between the State and the Main Line Corporations. At the second level, the legal nature of water rights for the Source Corporations was not explicitly defined although one could reasonably guess that it intended to be defined as usufruct right.

[173] Ibid, 72.

The latest development in the Eastern Route made an effort in incorporating the two separate levels of the project corporations, which made the Eastern Route Corporation the water abstraction right holder. In the case of the Middle Route, while there is no explicit change in the roles of the Middle Route Bureau, the 2014 Regulation on Administration of Water Supply and Use for the SNWDP provides a mechanism for the Middle Route to make the same integration. As discussed in the previous section on the Middle Route, the Middle Route Bureau can be authorised by the State Council as the Middle Route project management unit to hold the abstraction right.

In the context of a natural river basin, provincial governments are allocated water use quotas based on the river basin plans. Such water use quotas present as regional water rights. Although the legal nature of regional water rights remains unclear in the current legal system, they are tradable according to the 2016 Provisional Measures on Administration of Water Rights Trading. In the case of the SNWDP, however, the provincial level water right was arranged as water supply contracts, with lower security and weaker legal protection than regional water rights. The proposed Main Line Corporations within which provinces share, together with the State as the controlling shareholder, was not put in place. Instead, the newly established Eastern Route Corporation became a state-owned enterprise with the State Council as the single investor. This may be consistent with the arrangement of water supply contracts, which could be a compromised outcome as redistribution of interests experienced political difficulties. Nevertheless, the differentiation in the water rights arrangements at the provincial level between natural water systems and engineered project systems like the SNWDP lacks convincing rationale that is simply based on political compromise.

6.3.2 A New Proposal of Water Rights Arrangements for the Project

Bearing the drawbacks of the MWR proposal in mind, an alternative water rights arrangement for the SNWDP is proposed in this section. As shown in the following figure (Fig. 6.5), it involves five levels of organisations and corresponding water rights. The proposal acknowledges the potential challenge in the existing legal and policy systems that already put the contractual-based water rights arrangement in place for the SNWDP. Major reform will be needed to facilitate such an institutional change. To aim to nurture the world's potentially largest water rights market, this proposed framework attempts to move towards a water rights arrangement that provides greater certainty and security to right holders and allows higher tradability.

Standing at the second level in the framework, the Middle Route Bureau and the Eastern Route Corporation are conferred an overall water usufruct right originating from the state ownership at the first level. Specifically speaking, they are water abstraction rights in the form of water abstraction permits. As the core legal system

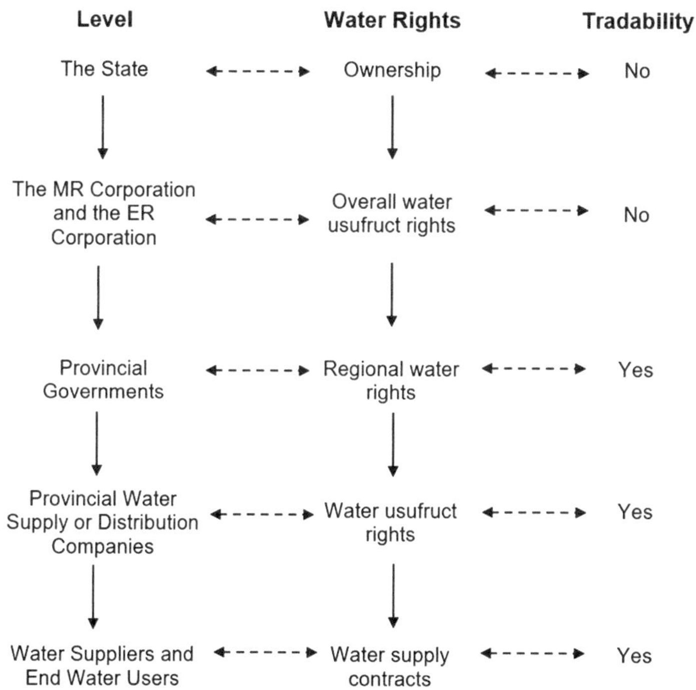

Fig. 6.5 Five level water rights arrangements for the South-to-North Water Diversion Project in China

pertaining to the water rights initialisation and arrangements, water abstraction permits should be incorporated into the institutional framework for the SNWDP. While water abstraction rights are generally tradable, restrictions may be necessary to the overall water usufruct rights for the SNWDP to ensure water availability for the water receiving regions.

At the third level, regional water rights are issued to provincial governments in the form of water use quotas defined by water allocation plans, and tradable according to the 2016 Provisional Measures on Administration of Water Rights Trading. In other words, provincial governments are allocated a share from the overall water abstraction right. The water resource from the SNWDP is treated here similarly to a natural common resource, such as a river or an aquifer. And a regional water right here is a province's share of the SNWDP water resource, and represents the water available for allocation within the province. Such regional water rights should be theoretically tradable between provinces to maximise effectiveness of the market allocation mechanism, and of course, subject to restriction rules for protection of third-party effects as well as the environment. The additional water resources from the SNWDP in the receiving regions should be integrated within water resources allocation plans. However, the formulation of water allocation plan is lagging behind the construction of water diversion projects.[174] As a result, the water allocation of inter-basin diversion in the water receiving regions is yet to be fully incorporated in the current water resources allocation plans.[175] Nevertheless, the provincial regional water rights can be further allocated to lower levels of government, or directly to water supply companies or other water users in the form of a water abstraction permit. Provincial water supply or distribution companies are issued a normal water abstraction right which again is tradable (represented as the fourth level in the framework). Annual allocations based on the water abstraction right should also be further unbundled and tradable on a temporary or seasonal basis. Provincial governments can establish restriction rules for the SNWDP related water rights trading to protect its local economy, community, and the environment. And finally, water supply contractors at the fifth level in the framework should similarly be allowed to trade their contractual water rights.

6.4 Concluding Remarks

China has been working hard to initialise water rights and make them tradable for more efficient and sustainable water allocation. With the philosophy of water use changed towards sustainability, private property rights to water began to be recognised by law and gradually known by the public. A process from water planning through water quantity allocation to water abstraction permits has divided usufruct rights from state ownership over water resources, without which water rights trading cannot exist. Under the traditional institutional framework of government allocation, the definition of tradable water rights in China is still in its infancy. The boundary of government involvement is yet to be clearly set, and water rights are legally weak in the sense that they have to be subject to significant government intervention. Based on the analysis of China's water abstraction rights and reflections on overseas experiences, this chapter has clarified several essential aspects of the legal framework. These clarifications include the following:

- Water rights arrangements are not only limited in usufruct rights. In many other cases, water supply contracts, particularly those of farmers are clearly specified, quantified, registered, and legally acknowledged so that they share a comparable level of certainty and security to property rights. Understanding this is important for China's water law and policy reform. As the largest water consumer of China, farmers should be given more secure water rights. This would not only augment efficiency of water rights trading but improve equitability of water allocation.

[174] China Ministry of Water Resources et al., *Water Entitlements and Trading Project, Phase 2 Final Report*, 2007, https://www.environment.gov.au/system/files/pages/d460a425-1223-401a-a544-cd4ee5916fc7/files/wet-phase2-report-english.pdf (accessed May 20, 2017), 65.

[175] Ibid.

- The nature of regional water rights at the local levels (provincial governments, prefectures, and counties) will have to be further clarified. Regional water rights are either allocated directly to water users in the form of a water abstraction permit, or to lower level governments, which will create a lower level regional water right. Therefore, there are three levels of regional water rights (provincial, prefecture, and county). Water allocation plans at the basin and regional levels need to be developed to specify regional water rights. When local governments participate in water rights trading, their roles as a water right holder need to be clearly separated from their public administration power.
- Some important values of water are yet to be recognised by the current legal systems of water rights. Water abstraction rights only cover consumptive water uses and leave instream (non-consumptive) uses out of legal protection as statutory water rights. Water rights trading is favoured and promoted in China's water reform not only for its function in allocation efficiency but also for its effectiveness in water use control and conservation. Environmental water rights should be recognised and accommodated in the legal structure of trading in water rights.

In response to the above-mentioned issues, the chapter has made recommendations for China's water reform by reflecting on overseas experiences. Proposals for both the overall water rights arrangement framework and some specific elements of water rights have been developed for the purpose of improving right security to facilitate the growth of water rights markets. Recommendations made throughout the chapter are summarised as below:

- It has been observed that statutory governance of water rights is the main approach to the development of water rights arrangements even in common law countries. Riparian rights and appropriative rights stemming from common law have been governed by statutes as well as administrative permit systems while other statutory water rights are developed to protect public values of water resources. In the context of China, two legal origins of formal water rights, namely permits and statutes, fail to accommodate many critical interests in water due to its limited coverage. Among those missing interests, irrigation water rights of WUAs and individual farmers, and environmental water rights are the ones that urge immediate policy response. They deserve to be explicitly recognised by statutes for their significance in water allocation. Furthermore, how irrigation water rights and environmental water rights can fit in the permits system or statutes for obtaining legal origins represents a critical reform issue for China.
- To better protect farmers' water rights for a more active trading market, farmers' water rights certificates should be recognised by statutes and universally implemented in irrigation districts. To achieve this, quantification of water rights at the levels of WUAs and individual farmers is required. By referring to those experiences in Australian jurisdictions such as Victoria and Queensland, China's irrigation water rights could be a three-level framework, within which WUAs stand in the middle to bridge irrigation authorities holding water abstraction rights, and individual farmers holding registered water rights in the form of clearly quantified water supply contracts.

- In the context of China, where a tradition of government control in water allocation has long been followed, the newly recognised water abstraction rights are deficient in certainty and security. To stimulate engagement in water trading markets, water abstraction rights in China need to be strengthened in several essential elements. More importantly, transferability and divisibility of the rights need to be further augmented by allowing temporary water rights transfers. Without these provisions, the functions of the market mechanism in resource allocation efficiency and flexibility are less likely to be maximised.
- Instream uses of water for the environment are not well protected in the current legal systems of water rights arrangements, although efforts have been made in the policy development such as inclusion of the environment in prioritisation of water abstraction permits. However, the over-allocated water resources of China can only be effectively relieved of heavy burden with stronger protection to environmental water rights.

Water rights arrangements for the SNWDP were also presented in detail in this chapter because of its significance in the establishment of China's water rights market. Based on a detailed analysis of the proposal developed by the MWR policy advisors, it is clear that water abstraction permits as China's key form of water usufructuary rights are yet to be fully incorporated into the institutional arrangements of the project. The chapter attempted to do so by proposing a new institutional framework for the project, in which four levels of water rights are established and given general tradability in order to facilitate market growth. With tradable water rights in place, China needs to further design a water rights administration system and trading rules for the water rights market. Indeed, how water rights are governed and administered has important implications for market performance. The next chapter addresses this issue.

References

Australia Bureau of Agricultural and Resource Economics and Sciences (ABARES). (2016). *Australian water markets report 2014–15*. Canberra: Department of Agriculture and Water Resources.

Australia Bureau of Statistics. *2004–05 Water access entitlements, allocations and trading*. http://www.abs.gov.au/ausstats/abs@.nsf/mf/4610.0.55.003. Accessed 28 Sept 2008.

Australia China Environment Development Program. (2006). *Report on the development of WATER rights and trading in the peoples' Republic of China*. Canberra: Department of Agriculture, Fisheries and Forestry of Australia.

Australia House of Representatives Standing Committee on Agriculture, Fisheries and Forestry. (2004). *Getting water right(s)-the future of rural Australia*. Canberra: Commonwealth of Australia.

Australia National Water Commission (NWC). (2011a). *Water markets in Australia: A short history*. Canberra: National Water Commission.

Australia National Water Commission (NWC). (2011b). *The national water initiative – Securing Australia's water future: 2011 assessment*. Canberra: National Water Commission.

Australia National Water Commission (NWC). (2011c). *Third biennial assessment of progress in implementation of the national water initiative*. http://www.agriculture.gov.au/water/policy/nwi. Accessed 30 Jan 2017.

Australia National Water Commission (NWC). (2013). *Australia water markets report 2012–13*. Canberra: National Water Commission.

Australia National Water Commission (NWC). *Australia's water blueprint: National reform assessment 2014*. http://webarchive.nla.gov.au/gov/20160615062247/http://www.nwc.gov.au/publications/topic/assessments/australias-water-blueprint-national-reform-assessment-2014. Accessed 18 May 2017.

Australia Productivity Commission. (2003). *Water rights arrangements in Australia and Overseas*. Melbourne: Productivity Commission.

Bartley, M., Fisher, D., & Queensland University of Technology. (2004). *Trading in water rights: Towards a national legal framework*. Sydney/Brisbane: Phillips Fox/Queensland University of Technology.

Bates, S. F., Getches, D. H., MacDonnell, L. J., & Wilkinson, C. F. (1993). *Searching out the headwaters: Changes and rediscovery in western water policy*. Washington, DC: Island Press.

Bauer, C. J. (2004). *Siren song: Chilean water law as a model for international reform*. Washington, DC: Resources for the Future.

China Ministry of Water Resources, AusAID and Australian Department of the Environment, Water, Heritage and the Arts. (2007). *Water entitlements and trading project, Phase 2 Final Report*. https://www.environment.gov.au/system/files/pages/d460a425-1223-401a-a544-cd4ee5916fc7/files/wet-phase2-report-english.pdf. Accessed 20 May 2017.

China Ministry of Water Resources, Development Research Centre. (2003). An introduction to research on the construction and administration system of the South-to-North project. *China Water Resources, 1B*, 72–74.

China Ministry of Water Resources, Water Management Department. (2005). *The Report on pilot experiences of establishing the water rights system part 1: Materials of water rights transfer in the inner mongolia autonomous region*. Beijing: China Water Publishing Co.

Cui, J. (2006). *Studies on quasi property rights* (2nd ed.). Beijing: Law Press.

Deason, J. P., Schad, T. M., & Sherk, G. W. (2001). Water policy in the United States: A perspective. *Water Policy, 3*, 175–192.

Getches, D. H. (1997). *Water law in a Nutshell* (3rd ed.). St. Paul: West Publishing.

Guy, D. J. (2008). A model water transfer act for California: An agricultural perspective. *Hastings West-Northwest Journal of Environmental Law & Policy, 14*, 709–724.

Hodgson, S. (2006). *Modern water rights: Theory and practices*. Rome: Food and Agriculture Organization of the United Nations.

Mckenzie, M. (2009). Water rights in NSW: Properly property? *Sydney Law Review, 31*, 443–463.

O'Donnell, M. (2013). The national water initiative, native title rights to water and the emergent recognition of indigenous specific commercial rights to water in Northern Australia. *The Australasian Journal of Natural Resources Law and Policy, 16*(1), 83–100.

Poirier, R., & Schartmueller, D. (2012). Indigenous water rights in Australia. *The Social Science Journal, 49*(3), 317–324.

Sax, J. L. (1990). *The constitution, property rights and the future of water law*. Boulder: Natural Resources Law Center, Western Water Policy Project Discussion Series Paper No.2.

Slater, S. S. (2005). "A prescription for fulfilling the promise of a robust water market". Symposium on the 25th anniversary of the report of the governor's commission to review California water rights law. *McGeorge Law Review, 36*, 253–294.

Teerink, J. R., & Nakashima, M. (1993). *Water allocation, rights and pricing: Examples from Japan and the United States* (World Bank Technical Paper, No. WTP 198).

Townsend, D., & Adams, O. (2016, September). Water rights and trading in Australia. *Australian Environmental Law Digest, 3*(2), 3–4.

Zhang, L., Xu, Z., Guo, P., & Zhang, Y. (2007). A review of the Australian experiences in the water quality security and administration of the snowy mountains project. *South-to-North and Water Conservancy Technology 2*. http://www.hwcc.com.cn/newsdisplay/newsdisplay.asp?Id=178716. Accessed 28 Sept 2008.

Legal and Policy Sources

China Ministry of Water Resources Notices on the Implementation of the Regulation on Administration of Water Abstraction Permits and Water Resources Fee Collection. (2006). http://www.gdwater.gov.cn/cms/yewuzhuanji/szygl/ggtz/20060703104256.html. Accessed 28 Sept 2008.
Constitution of the People's Republic of China (2004 Amendments). (1982). http://www.npc.gov.cn/englishnpc/Constitution/node_2825.htm. Accessed 21 May 2017.
Implementing Measures on the Water Abstraction Permits System (China State Council). (1993). http://www.chinawater.net.cn/law/W03.htm. Accessed 28 Sept 2008.
Implementing Rules on Water Abstraction Permits in the Yellow River Basin (Yellow River Conservancy Commission). (1994). http://huanghe.ikdo.com.cn:81/mycms/zcfg/20988.jhtml. Accessed 21 May 2017.
Measures on Administration of Water Abstraction Permits (China Ministry of Water Resources). (2008). http://www.gov.cn/flfg/2008-04/17/content_947055.htm. Accessed 21 May 2017.
Measures on Implementation of the Water Law of People's Republic of China for Beijing Municipality (Standing Committee of Beijing Municipal People's Congress). (2004). http://www.bjcg.gov.cn/lawsearch.bbscs?action=read&idStr=119. Accessed 28 Sept 2008.
Measures on Supervision and Administration of Water Abstraction Permits (China Ministry of Water Resources). (1996). http://www.hwcc.com.cn/newsdisplay/NewsDisplay.asp?id=46093. Accessed 26 June 2008.
Property Law of the People's Republic of China. (2007). http://www.npc.gov.cn/englishnpc/Law/2009-02/20/content_1471118.htm. Accessed 21 May 2017.
Provisional Measures on Administration of Water Rights Trading (China Ministry of Water Resources). (2016). http://www.gov.cn/zhengce/2016-05/22/content_5075679.htm. Accessed 27 May 2017.
Provisions on Application and Approval Procedures for Water Abstraction Permits (China Ministry of Water Resources). (1994). http://www.mwr.gov.cn/zcfg/19940609/13845.asp. Accessed 26 June 2008.
Regulation on Administration of Water Abstraction Permits and Water Resources Fee Collection (China State Council). (2006). http://www.gov.cn/zwgk/2006-03/06/content_220023.htm. Accessed 21 May 2017.
Regulation on Administration of Water Supply and Use for the South-to-North Diversion Project (China State Council). (2014). http://en.pkulaw.cn/display.aspx?cgid=219668&lib=law. Accessed 21 May 2017.
Regulation on Water Conservation for Tianjin Municipality (2005 Amendments) (Standing Committee of Tianjin Municipal People's Congress) (2002). http://www.chneef.com/law/qyfg/200504/20050423054233.html. Accessed 28 Sept 2008.
Water Law of the People's Republic of China. (1988) http://english.mofcom.gov.cn/article/laws-data/chineselaw/200211/20021100053796.shtml. Accessed 21 May 2017.
Water Law of the People's Republic of China. (2002). http://www.npc.gov.cn/englishnpc/Law/2007-12/12/content_1383920.htm. Accessed 21 May 2017.

Websites

The official website of the Construction and Administration Bureau of South-to-North Water Diversion Middle Route Project. http://www.nsbd.cn/n524903/index.html. Accessed 27 Aug 2008.
The official website of the Han River Group of Water Conservancy and Hydroelectric Power. http://www.hjgrp.com/about_us/lsyg_yg.asp. Accessed 28 Sept 2008.

The official website of the Henan Section of the South-to-North Project. http://www.hnnsbd.gov.cn/index/gaikuang.asp. Accessed 28 Sept 2008.
The official website of the South-to-North Water Diversion Project. http://www.nsbd.gov.cn/. Accessed 28 Sept 2008.
The official website of the Water Source Corporation of the South-to-North Middle Route. http://www.zxsygs.com/html/gsgk/. Accessed 27 Aug 2008.

Chapter 7
Administering Water Rights and Trading

Abstract From an economic perspective, water rights administration is an important variable of transaction costs for water rights trading. Globally in many countries in favour of market allocation mechanisms, water law reforms are attempting to remove the institutional barriers to their water trading markets. China faces a similar but more daunting challenge given that its water rights administration is to be established virtually from scratch. In other words, the current system lacks fundamental elements for water rights trading administration, and yet without it, a formal water market cannot occur. This chapter reviews international experiences from which China may learn to overcome the essential deficiencies in its existing legal system in water rights administration. Four major components regarding water rights administration highlighted in this chapter include centralisation of water rights administration, the water rights titling/registration system, provision of trading platforms, and market regulation and third-party protection. Reform recommendations will be made for China to address those critical components in water rights administration.

Keywords Water rights administration • Water rights titling/registration • Water trading platforms • Water market regulation • Third party protection

7.1 The Current System of China's Water Rights Administration

7.1.1 Hierarchy of Water Rights Administration

As described in Chap. 3, China's water legislation has established a system of water resources administration called the Competent Department of Water Administration (CDWA), which involves multiple levels of governmental organisations.[1] The MWR directly under the State Council is the top water authority of China in charge of the unified administration and supervision of the state's water resources. It has a range of responsibilities in water policy development, water resources allocation and

[1] For more details about the CDWA, refer to Sect. 3.2.1.1.

M. Jiang, *Towards Tradable Water Rights*, Global Issues in Water Policy 18,
DOI 10.1007/978-3-319-67087-4_7

administration.[2] Seven river/lake basin authorities exercise water administration and supervision in their respective boundaries as the MWR agencies. At the local levels, above or at the county level, WRBs/WABs are responsible for the unified management and supervision of water resources within their corresponding administrative units as counterparts of the MWR.[3]

Similarly, water rights have been administered in China in light of this hierarchic system. The power of water rights administration has been divided among various levels of governmental departments mainly based on the 2006 Regulation on Administration of Water Abstraction Permits and Water Resources Fee Collection and the 2008 Measure on Administration of Water Abstraction Permits. The division of power across the government levels is as follows:

- The MWR is responsible for organising the implementation of the water abstraction permits system, and supervising and administering the system across the entire state.[4]
- River/lake basin authorities under the MWR are responsible for organising the implementation of the water abstraction permits system, and supervising and administering the system within their jurisdiction in accordance with laws and regulations, or authorisation by the MWR.[5]
- Water administration departments above or at the county level are responsible for organising the implementation of the water abstraction permits system, and supervising and administering the system in corresponding administrative units in accordance with their hierarchical jurisdiction authorised by provincial governments (provinces, autonomous regions, or municipalities directly under the State Council).[6]

7.1.2 *Division of Power in Issuing Water Abstraction Rights*

The power of examining and approving water abstraction permits has been divided among different levels of authorities. The 1993 Implementing Measures on the Water Abstraction Permits System, in its Article 19, listed several categories of water abstraction that must be issued by the MWR or river basin authorities subject to the MWR authorisation. In 1994 and 1995, the MWR authorised the seven river/lake basin authorities to issue water abstraction permits within their jurisdiction. As

[2] For more details, see Sect. 3.2.1.1.

[3] For more details, see Sect. 3.2.1.1.

[4] 2008 Measures on Administration of Water Abstraction Permits (China Ministry of Water Resources), http://www.gov.cn/flfg/2008-04/17/content_947055.htm (accessed May 21, 2017), art. 3.

[5] Ibid. Also 2006 Regulation on Administration of Water Abstraction Permits and Water Resources Fee Collection (China State Council), http://www.gov.cn/zwgk/2006-03/06/content_220023.htm (accessed May, 21 2017), art. 3.

[6] Ibid.

a result, the MWR itself is no longer in charge of direct administration on the permits issuance. Currently, the power of examining and approving water abstraction permits is shared between river basin authorities and local departments for water administration (WRBs or WABs) at different levels (above or at county levels). River basin authorities are supported to be responsible for administering water abstraction with national importance. The 2006 Regulation provides a list of categories of water abstraction belonging to the jurisdiction of river basin authorities as follows.[7] Water abstraction permits other than those listed below are examined and approved by county and above WRBs or WABs in accordance with authorisation by their provincial government.[8] The water abstraction categories under river basin authorities' jurisdiction include the following:

- Water abstraction with an amount exceeding the given norm from the main stream of Yangtze River, Yellow River, Huai River, Hai River, Luan River, Pearl River, Songhua River, Liao River, Jinsha River, and Han River; from Taihu Lake; or from other appointed courses of trans-provincial rivers or lakes.
- Water abstraction with an amount exceeding the given norm from the appointed courses of transboundary rivers and international border rivers.
- Water abstraction with an amount exceeding the given norm from provincial border rivers or lakes.
- Water abstraction involving two or more provinces, autonomous regions or municipalities directly under the State Council.
- Water abstraction for major construction projects examined and approved by the State Council or the department in charge of finance under the State Council.
- Water abstraction from river or lake courses or sections directly managed by river basin authorities.

WRBs or WABs at the provincial level play an advisory role in the examination and approval of water abstraction permits under the jurisdiction of river basin authorities, although they do not have power to issue permits to the above listed water abstractions. The applications are submitted to the provincial WRB or WAB where the site for water abstraction is located.[9] The WRB or WAB will then provide the river basin authority with all the documentation including primary advice on the proposed water amount for abstraction, requirements on water quality for abstraction and return flow, the total amount of water already issued for abstraction from the water system under application, and the situation of water quality in the water function zone.[10]

The power to examine and approve applications for groundwater extraction permits in the area subject to groundwater utilisation restriction remains with

[7] Ibid, art.14.

[8] Ibid.

[9] 2008 Measures on Administration of Water Abstraction Permits, art. 11. And 2006 Regulation on Administration of Water Abstraction Permits and Water Resources Fee Collection, art.10.

[10] 2008 Measures on Administration of Water Abstraction Permits, art.12.

higher-level authorities. The 2008 Measures separate the power between provincial WRBs or WABs and river basin authorities:

> In the case that the application is raised for extracting groundwater in the area with groundwater utilisation restriction, the provincial WRB or WAB where the site of extraction is located is responsible for examining and approving the application. Among those applications, however, if the groundwater extraction is requested for major construction projects that are examined and approved by the State Council or the department of finance under the State Council, the river basin authority is in charge of examining and approving such application.[11]

7.1.3 *Issuance of Water Abstraction Permits*

7.1.3.1 Applications

The applicant who pursues a water abstraction permit is required to submit the application with supporting documents which provide adequate information for the relevant authority to make the assessment. A standard application form for water abstraction permits is designed by the MWR and used across China. Information included in the application form consists of[12]:

- personal details about the applicant, including name, title and address;
- justification of applying for water abstraction;
- duration of water abstraction;
- purpose and volume of water requested for abstraction, and the amount of water use for each month of the year;
- name of water source and location from which water is to be abstracted; water abstraction pattern, water metering method, and water saving measures;
- information regarding return flows such as location, major pollutants, and wastewater treatment measures.

The applicant is also required to provide a statement that describes the impact of the proposed water abstraction on third parties.[13] In the case where the water abstraction may potentially result in harm to a third party, the applicant needs to obtain a consent or agreement with the third party regarding the issue.[14]

Applications for water abstraction for construction projects require preparation of a water resources impact assessment to determine whether the water source as well as the requested water use is reasonable, and whether any ecological and environmental impacts may result from such water use.[15] The assessment has to be

[11] Ibid, art.16.

[12] 2006 Regulation on Administration of Water Abstraction Permits and Water Resources Fee Collection, art.12.

[13] Ibid.

[14] Ibid, art. 11; And 2008 Measures on Administration of Water Abstraction Permits, art.10.

[15] 2006 Regulation on Administration of Water Abstraction Permits and Water Resources Fee Collection, art.11.

formulated by a qualified entity with exception of those construction projects that abstract a small amount of water and thus are expected to have little impact on the surrounding environment. Rather than getting a qualified entity to undertake the assessment, the applicants only need to fill in a standard justification form to address the issue of the water resources impacts.[16]

When the need exists for a new sewage outlet for return flows, the water abstraction permit application should also include an application for the sewage outlet setup.[17] For water abstractions of which return flows use an existing sewage outlet, the application should be submitted with consent from the department of water administration at or above county levels or the responsible river basin authority.[18]

7.1.3.2 Consultation and Examination

Public consultation is not a compulsory process for water abstraction permit applications in China. Public announcements and hearings can apply if in the opinion of the responsible authority, the water abstraction under application involves social public interests.[19] In addition, if major effects on third parties are associated with the approval, the authority is required to inform both the applicant and the interested party before it makes any decision.[20] Public hearings may then be organised by the authority upon request by the applicant or the interested party.[21]

In some cases, governmental consultation is required for processing a water abstraction permit application. For example, the 2006 Regulation in its Article 19 states that the urban construction department should be consulted about applications that request groundwater extraction in the urban planning areas. The relevant WRB/WAB or river basin authority in the area where the abstraction site is located may also be consulted during the assessment process if more information is needed from their perspective.[22] Furthermore, technical consultation has to be arranged as well for the water resources impact assessment for construction projects based on which the application is to be assessed.[23]

Applications are examined on the basis of information provided by applicants as well as consultations with stakeholders concerned. The responsible authority is required to make a decision of denial in the following circumstances[24]:

---

[16] 2008 Measures on Administration of Water Abstraction Permits, art. 8.

[17] Ibid, art. 1.

[18] Ibid, art. 10.

[19] 2006 Regulation on Administration of Water Abstraction Permits and Water Resources Fee Collection, art. 18.

[20] Ibid.

[21] Ibid.

[22] 2008 Measures on Administration of Water Abstraction Permits, art. 19.

[23] Ibid, art. 9.

[24] Ibid, art. 20. And 2006 Regulation on Administration of Water Abstraction Permits and Water Resources Fee Collection, art. 20.

- Groundwater extraction in a forbidden area.
- Excessive amount of water is requested for abstraction in an area where the total amount of water abstraction has already reached its total control quota.
- Water abstraction likely to result in major damages to water function zones. For example, water quality falls below water quality standards for the function zones as a result of water quantity reduced for the abstraction.
- Unreasonable locations for abstraction or return flows, including (1) the sewage outlet for return flows is setup in a reserve area for protection of drinking water sources; (2) concentration of major pollutants in return flows exceeds national or local disposal standards; (3) water quality of the water area receiving return flows likely to fall below the water quality standard for water function zones; (4) return flows not meeting requirements of the total volume control of wastewater discharge; (5) return flows not meeting requirements of groundwater replenishment.
- Construction projects use their own equipment to extract groundwater in the case that their need for water can be satisfied by the public network of urban water supply.
- Serious damage to third parties or social public interests is likely to result from the water abstraction.
- The projects that are required for submission for record fail to do so.

7.1.3.3 Decision Notification and Registration

Once a decision of approval on an application is made, procedures for notifying the approval are employed. As discussed in Chap. 6, the applicant whose application has been approved will receive an approval document based on which the construction of water infrastructure is allowed to begin.[25] After the construction work has been completed and has passed examination, the applicant will be officially issued a water abstraction permit.[26] The applicant is informed of the decision in writing with other related information, such as the responsible authority in charge of supervision and administration of the permit and matters pertaining to collection and administration of water resources fees.[27] The decision of approval also requires notification to the water administration department at the county level in the region where the abstraction site is located.[28]

According to the 2003 Administrative Permission Law, the decisions of approval on an administrative permit application are accessible to the public.[29] The 2008

[25] For further details, refer to Sect. 6.2.2.

[26] See Sect. 6.2.2.

[27] 2008 Measures on Administration of Water Abstraction Permits, art. 26.

[28] 2006 Regulation on Administration of Water Abstraction Permits and Water Resources Fee Collection, art. 23.

[29] 2003 Administrative Permission Law of the People's Republic of China, http://www.npc.gov.cn/englishnpc/Law/2007-12/05/content_1381966.htm (accessed May 22, 2017), art. 40.

Measures on Administration of Water Abstraction Permits specifically requires decisions of issuing water abstraction permits to be announced publicly on an annual basis.[30]

Water abstraction permits are registered by the responsible authority in charge of examining and approving the permit applications. A standard registration form designed by the MWR is universally used across China. Authorities are required to annually report their issuance of permits to their next higher counterparts for record.[31] Provincial water authorities are responsible to collect registries of water abstraction permits and other administrative records, including the total amount of water approved for abstraction within their respective administrative unit and report annually to river basin authorities.[32] Based on the information provided by the provincial water authorities, river basin authorities then build up divisional registries in terms of different water systems in the river basin.[33] It is also required that river basin authorities report to the MWR information related to water abstraction approval and permits issuance in the river basin on an annual basis.[34]

7.1.3.4 Appeals Against Decisions

With an appeals mechanism in place, administrative decisions in China can be reviewed in the forms of administrative reconsideration or judicial lawsuit to protect individuals' legal rights against government. The Administrative Reconsideration Law of 1999 enables individuals including citizens, legal persons and other organisations to apply for an administrative reconsideration if they believe that their applications for an administrative permit or licence have met legal requirements, but the administrative authority has failed to process their applications in accordance with law.[35] If the decision for an administrative reconsideration cannot satisfy the applicant, they may bring an administrative lawsuit before a People's court.[36] Alternatively, based on the Administrative Procedure Law 1989, the applicant may directly bring a suit rather than going through the procedures of administrative reconsideration before doing so.[37]

---

[30] 2008 Measures on Administration of Water Abstraction Permits, art. 31.

[31] 2006 Regulation on Administration of Water Abstraction Permits and Water Resources Fee Collection, art. 46.

[32] 2008 Measures on Administration of Water Abstraction Permits, art. 47.

[33] Ibid.

[34] Ibid.

[35] 1999 Law of People's Republic of China on Administrative Reconsideration, http://www.npc.gov.cn/englishnpc/Law/2007-12/11/content_1383562.htm (accessed May 22, 2017), art. 6.

[36] Ibid, art. 5. It should be noted that some decisions on administrative reconsiderations are terminal by law and therefore cannot be brought to a court.

[37] 1989 Administrative Procedure Law of People's Republic of China, http://www.npc.gov.cn/englishnpc/Law/2007-12/12/content_1383912.htm (accessed May 22, 2017), art. 2, art. 11.

Decisions on water abstraction permit applications are subject to the above-mentioned provisions. The specific regulations related to the water abstraction permit system also identify the options for applicants to appeal decisions. However, there are some changes made to address the appeal issue between the 1993 Implementing Measures on the Water Abstraction Permits System and the new provisions. The 1993 Implementing Measures explicitly allowed applicants to appeal decisions of denial on their applications by stating in Article 20:

> When the application for a water abstraction permit has been objected, the applicant, who believes that his or her application has satisfied legal requirements, may apply for an administrative reconsideration or bring a lawsuit before a people's court in accordance with law.[38]

Consistent with this regulation, most implementing measures on water abstract permits at local levels included a similar provision to allow applicants to appeal a decision of denial on a permit application. Interestingly, however, this provision has not been reinstated in either the 2006 Regulation on Administration of Water Abstraction Permits or the 2008 Measures on Administration of Water Abstraction Permits at the state level. Instead, an early legal document in 1999, entitled "the MWR Interim Rules on Administrative Reconsideration", governs administrative reconsiderations at the MWR level and the river basin authority level.[39] According to this regulation, applicants who apply to the MWR, river basin authorities or provincial water authorities for a water abstraction permit may pursue an administrative reconsideration if they believe that their permit applications have met legal requirements, but failed to be processed in accordance with law.[40]

It can be argued that water authorities' decisions on water abstraction permits are generally subject to administrative and judicial review on appeal, although only applicants themselves, other than other third parties who may be potentially affected, have the right to appeal. As the special legislation directly related to the water abstraction permit system, neither the 2006 Regulation nor the 2008 Measures have included a provision recognising such right. Therefore, applicants have to seek legal foundations from other sources to support their appeal, which in most cases are additionally difficult.

7.1.4 Modification and Transfer of Water Abstraction Permits

To a large extent, modification of water abstraction permits is generally restricted in China. As discussed in Chap. 6, permits cannot be modified but must be reapplied for if modification to existing water abstraction rights involve changes in the

[38] 1993 Implementing Measures on the Water Abstraction Permits System (China State Council), http://www.chinawater.net.cn/law/W03.htm (accessed September 28, 2008), art. 20.

[39] 1999 MWR Interim Rules on Administrative Reconsideration (China Ministry of Water Resources), http://www.sxwater.gov.cn/oldweb/zcfg/SLXZCFFLFG/SLXZCFFLFG1019.htm (accessed September 28, 2008), art. 2.

[40] Ibid, art. 5.

following aspects: the volume and purpose of use, the water source, the point of abstraction or the return flows, the quantity or pattern of return flows, and major pollutants in return flows or the wastewater treatment method.[41] It means that a permit holder may lose their permit if any one of these changes takes place. As an exception, however, the modification involving a change in the volume of use is allowed if the change is a result of a water abstraction right transfer.[42] In addition, the 2008 Measures on Administration of Water Abstraction Permits allow modifications such as changes in the name or title of the abstractor and changes resulting from water abstraction right transfers.[43]

Since the current practices in water right transfer are still limited in China, the relevant procedures to administer water right transfers remain in their infancy. Essential procedures and rules on water rights transfers including applications, assessment and approval, and registration were not in place until 2016 when the Provisional Measures on Administration of Water Rights Trading were developed. During the 10 years between 2006 and 2016, although conservation-based water trading was encouraged by the central government, piloting practices mostly took place in the Yellow River Basin. The Trial Implementing Measures on Administration of Water Rights Transfer in the Yellow River Basin were developed as early as 2004. As examined in Chap. 4, water rights transfer in the Yellow River Basin is conservation based. An industrial project invests in construction of water conservation rehabilitation projects in an irrigation district and acquires the portion of water under the irrigation district's water abstraction permit saved from irrigation through the investment.[44] The water rights transfer practices in the Yellow River Basin were predominantly driven by government, and thereby local governments, river basin authorities, and local water authorities were largely involved during the process.

According to the 2004 Trial Implementing Measures, inter-provincial water rights transfers are not allowed in the Yellow River Basin.[45] An overall water rights transfer plan for a provincial region is required to implement water rights transfer and have it approved by the YRCC.[46] The provincial water administration department (WRB or WAB) is responsible to formulate such plan in consultation with the development and planning department at the same level.[47] Proposals of water rights transfers in the short term are included in the plan, and each one has to identify transferee and transferor, and the proposed amount of water for transfer.[48] The

---

[41] See Sect. 6.2.3.

[42] See Sect. 6.2.3.

[43] 2008 Measures on Administration of Water Abstraction Permits, art. 28.

[44] For further information about the water rights transfers in the Yellow River Basin, refer to Sect. 4.4.3.

[45] 2004 Trial Implementing Measure on Administration of Water Rights Transfer in the Yellow River Basin (Yellow River Conservancy Commission), http://www.yrcc.gov.cn/ziliao/zcfg/guizhang/200612/t20061222_9390.htm (accessed September 26, 2008), art. 3.

[46] Ibid, art. 3.

[47] Ibid, art. 7.

[48] Ibid, art. 14.

power of administrative approval on a water rights transfer application is divided between the YRCC and provincial WRBs or WABs. As Article 6 of the Trial Implementing Measures states,

> The YRCC is responsible for examining and approving water rights transfers on water abstraction permits issued by itself according to its jurisdiction, and water rights transfers in provinces (autonomous regions or municipalities directly under the State Council) whose water use quota has been exhausted. Other water rights transfers remain in the jurisdiction of provincial WRBs or WABs.[49]

Both parties involved in the transfer jointly submit an application to the provincial WRB or WAB which is responsible to make a preliminary assessment subject to the YRCC's jurisdiction.[50] The following documents are required for a water rights transfer application[51]:

- A copy of the water abstraction permit
- The proposed agreement of the water rights transfer signed by two parties
- The water resources assessment report for the proposed construction project
- The feasibility report for the water rights transfer
- Endorsement by the local government to whom the initial water right belongs.

Upon approval of the application, the provincial WRB or WAB then makes arrangements for the transferee and the transferor to come to a formal agreement for the transfer and formulate an implementation blueprint.[52] The water conservation project under the transfer is then to be constructed and completed subject to inspection by the YRCC and the provincial water administration department. Where all conditions are met, the transferee will apply for a new water abstraction permit and the transferor for a modification of their existing permit about the change in the amount of water on the permit.[53]

Fed on the piloting practices around the country, the 2016 Provisional Measures on Administration of Water Rights Trading have established procedures and rules on water rights trading, which provide that the MWR and its river basin authorities, together with county and above departments of water administration, are responsible for supervision and administration of water rights trading within their respective jurisdictions.[54] Articles 14–17 provide a four step procedure for trading in water abstraction rights as below:

Step 1: Prospective transferor applying to the original permit issuing authority for approval

[49] Ibid, art. 6.

[50] Ibid, art. 8, art. 9.

[51] Ibid, art. 8.

[52] Ibid, art. 20.

[53] Ibid, art. 23.

[54] 2016 Provisional Measures on Administration of Water Rights Trading (China Ministry of Water Resources), http://www.gov.cn/zhengce/2016-05/22/content_5075679.htm (accessed May 27, 2017), art. 4.

Step 2: The authority examining and approving the application
Step 3: Prospective transferor and transferee sign the trading agreement
Step 4: Prospective transferor and transferee applying for a new water abstraction permit or for modifications to the existing permit.

7.1.5 Loss of Water Abstraction Rights

A water abstraction permit can be cancelled by the issuing authority if the permit holder has stopped abstracting water for a consecutive period of 2 years.[55] However, the water abstraction permit can be kept if the non-abstraction is a result of *force majeure* or major technical transformation and the duration of the permit has not yet expired after 2 years of non-abstraction.[56] This provision reflects, in essence, the rule of beneficial use in the western US, where the appropriative water rights can be lost by failing to put water to beneficial use. Water abstraction permits are also subject to several revocation provisions as liabilities that permit holders take for not complying with related water legislation. A permit can be revoked in any one of the following circumstances[57]:

- The permit holder has abstracted water without approval and the circumstance is serious
- The permit holder has failed to abstract water in accordance with conditions prescribed by the permit and the circumstance is serious
- The permit holder has refused to implement an administration decision that the authority in charge has made to restrict the amount of water for abstraction, and the permit holder has refused to correct their illegal acts by the due date or the circumstance is serious
- The permit holder has transferred their water abstraction rights without approval, and refused to correct their illegal acts by the due date or the circumstance is serious
- The permit holder has conducted any one of the following illegal acts, and the circumstance is serious: failing to submit their annual water abstraction report, refusing to accept supervision and inspection or making falsification, or failing to meet the quality requirements for return flows
- The permit holder has failed to install water metering equipment, and the circumstance is serious
- The permit holder has failed to meet the requirement for water metering equipment, or properly operate the equipment, and the circumstance is serious.

[55] 2006 Regulation on Administration of Water Abstraction Permits and Water Resources Fee Collection, art. 44. And 2008 Measures on Administration of Water Abstraction Permits, art. 30.

[56] 2008 Measures on Administration of Water Abstraction Permits, art. 30.

[57] 2006 Regulation on Administration of Water Abstraction Permits and Water Resources Fee Collection, art. 51, art. 52, art. 53.

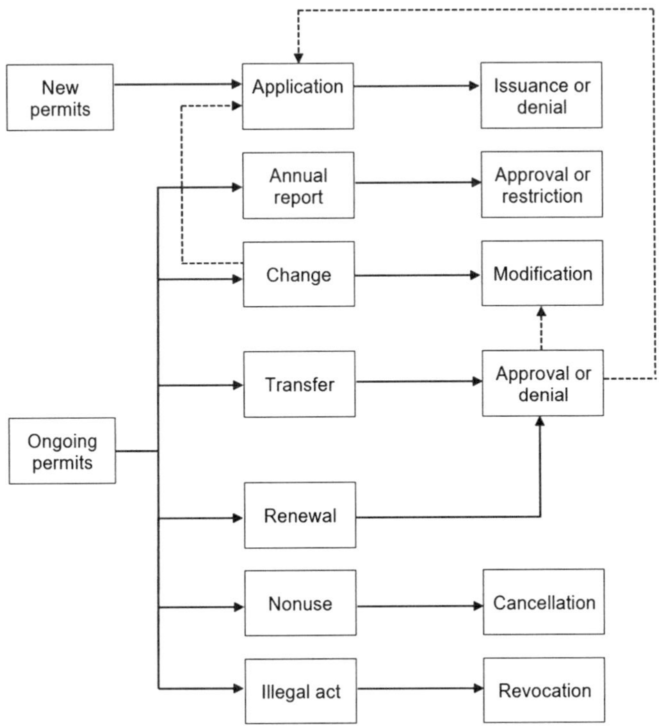

Fig. 7.1 The water rights administration framework in China

Administrative decisions on the cancellation and revocation of water abstraction permits are subject to review. Permit holders are able to request either an administrative reconsideration to higher level water authorities, or bring a lawsuit to a court if they are not satisfied with the decisions.

China's water rights administration framework is summarised in the above figure (Fig. 7.1), which demonstrates interactions between water users and water authorities in the processes of water rights administration in China.

7.2 Deficiencies in the Current System and Implications for Reform

The observations on the current water rights administration system in China have indicated strong government control as a remarkable feature. Governmental authorities issue new permits for water abstraction, monitor and restrict the exercise of water abstraction rights, and revoke existing permits to punish illegal acts in water use. Water rights holders, on the other hand, have little flexibility in changing their water use, whether they want to make modifications for their own use, such as put water to another purpose or location, or whether they want to alienate their water rights to others.

Changes, except those in permit holders' names and amounts of water as a result of water right transfer bring a requirement for new permit application, which means that the permit holder is unable to keep their water abstraction right as long as the essential conditions of water use under the right are changed. Transfers of water abstraction rights are mainly limited in conservation-based transfers subject to administrative control, no matter whether the transfer is accompanied with a change in the water use (such as a different location or purpose of use) or not. Tight control in water abstraction right trading is likely due to the fact that China's water reform begins from an administrative allocation approach with complete prohibition to transfer water abstraction permits. Free trading markets for water rights might have performed well in other jurisdictions around the world, but they require sufficient support by sophisticated institutional arrangements in association with water property rights as well as mature market rules. Unfortunately, the current system in China is yet to provide such conditions. China's approach to water reform, and the implementation of the market allocation mechanisms in particular, is learning by doing. Water law and policy in water rights trading has been gradually building up through reflections on the pilot practices. Current systems have been reformed to further facilitate wider application of water rights trading.

Water transfer from agriculture to industry has been a major trend in China's water use. Agriculture is required to conserve water by improving water use efficiency to satisfy new water demands in industry. Such water rights transfer represents a key type of water rights trading in China at the moment. Starting from this inter-sectoral permanent water rights transfer, China's water rights administration system will have to carefully address the critical issue of externalities. To this end, it is reasonable for the current system to enforce strict administrative approval for water rights transfers.

In contrast to the dominant administration power, the involvement of the judiciary in water rights administration remains limited. This is consistent with the fact that China's water rights are substantially subject to the power of governmental intervention. According to international experiences from other jurisdictions, the role of courts in administering water rights is largely dependent on the extent to which water rights are effectively conferred upon individual water users.[58] As a distinct example, in Colorado in the US, water courts are responsible for administering water rights with functions similar to those permitting authorities in other jurisdictions.[59] By comparison, the role of China's courts in water rights administration is generally limited to hearing appeals against related administrative decisions made by governmental agencies.

---

[58] Australian Productivity Commission, *Water Rights Arrangements in Australia and Overseas*, 2003, Melbourne: Productivity Commission, 75.

[59] David H. Getches, *Water Law in a Nutshell*, 3rd ed. (St. Paul, Minnesota: West Publishing Co., 1997), 154.

As observed earlier, the administration system for water rights trading remains underdeveloped within the current legal and institutional framework in China despite a growing body of provisions building up based on the Yellow River Basin experience. The review of the procedures of water rights transfer in the Yellow River Basin has shown that water transfers in the Yellow River Basin were mainly government controlled and regulated throughout the process as a result of heavy reliance on public administration by local governments, river basin authorities, and local water authorities. The following figure (Fig. 7.2) illustrates the stakeholder interactions involved in the first case of water rights transfer in Inner Mongolia as an example.

A transfer of 20.43 million cubic metres of water was made between the Irrigation District of the Southern Bank of Yellow River (thereafter the IDSB) and Inner Mongolia Electric Power Corporation (IMEPC).[60] As the transferor, the Irrigation District was represented by two relevant bureaus, namely Inner Mongolia Bureau of Yellow River Project Management (thereafter BYRPM) and Ordos Bureau of Irrigation Management of the Southern Bank of Yellow River, Inner Mongolia (thereafter OBIM).[61] The administration process of the water transfer involved three levels of water authorities: the YRCC (river basin authority), IMDWR (the provincial water department), and the Ordos WAB (the prefecture water bureau). BYRPM and OBIM on the transferor side and IMEPC on the transferee side jointly submitted the application to IMBWR who was responsible for a primary assessment. After that, the application was forwarded to the YRCC for final approval. Based on the approved application, two parties came into an agreement coordinated by the Ordos WAB who was also responsible for managing the construction of water saving projects involved in the agreement. Once the projects passed the inspection by both the YRCC and IMDWR, two parties then applied for modification of the water abstraction permit (for the transferor) and issuance of a new permit (for the transferee).

As the water supplier for the irrigation district, BYRPM was the water abstraction right holder under which an annual amount of 410 million cubic metres of water could be abstracted for the irrigation district.[62] It was also an agency for IMDWR, who was in charge of examining and approving water abstraction permits according to its jurisdiction.[63] The other stakeholder on the transferor side, OBIM was responsible for management, operation, and maintenance of water conservancy projects in the irrigation district.[64]

[60] China Ministry of Water Resources, Water Management Department. ed. *The Report on Pilot Experiences of Establishing the Water Rights System Part 1: Materials of Water Rights Transfer in the Inner Mongolia Autonomous Region* (Beijing: China Water Publishing Co., 2005), 113.

[61] Ibid, 227.

[62] Ibid, 130, 164.

[63] Information from the official website of the Inner Mongolia Department of Water Resources, http://www.nmgslw.gov.cn/info/infoView.jsp?idcontent=9 (accessed July 14, 2008).

[64] Water Management Department. ed. *The Report on Pilot Experiences of Establishing the Water Rights System Part 1*, 202.

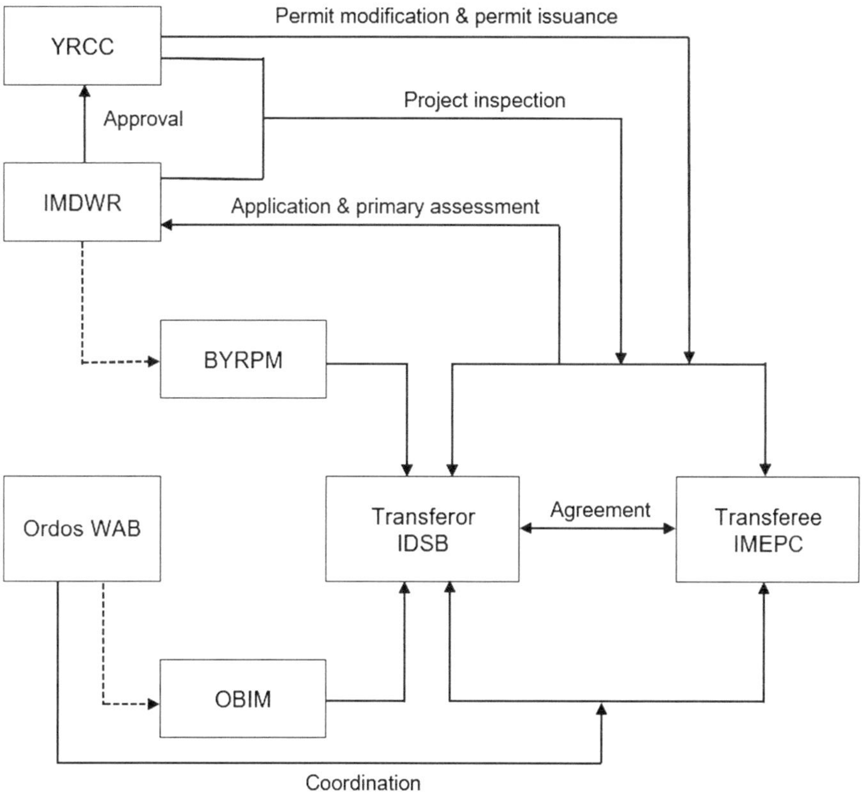

Fig. 7.2 Stakeholder interactions in water rights transfer in Inner Mongolia

As demonstrated in this case, water authorities not only administer but also directly perform water rights transfers in the Yellow River Basin. One may argue that BYRPM to some extent was independent of IMDWR as one of its divisional and subordinate organisations. However, they were closely related in terms of sharing coincident interests. Furthermore, BYRPM was also subject to the administrative leadership by IMDWR. These have explained why the public viewed IMDWR as a party in the water transfer agreement, instead of BYRPM. Such arrangement implied significant potential conflicts of interest. The provincial water administration department shares the responsibility for formulating the water rights transfer plan together with the development and planning department, which estimates and determines water quantity available for water rights transfer through water saving. Given the direct interest in water transfers, the water department may have incentive to overestimate water availability.

Due to the heavy reliance on government, pilot water rights transfers in the Yellow River Basin could even be viewed as a newer type of administrative transfer than market transfer. The substantial difference between the pilot transfers and the traditional administrative transfers is whether the transferor would be compensated for the alienation of the rights. Except for this differentiation, water transfers practised in the Yellow River Basin were government driven, administered and coordinated by government, and even performed by government. Transferees in these water rights transfers could neither choose with whom they would like to deal nor negotiate or bargain the prices. Indeed, water transfers were proposed by government in the form of an overall water rights transfer plan. With priority given to local economic development, the third-party effects of water rights transfer from agriculture to industry were given little attention. Ideally the agricultural community may not be negatively affected since only the portion of conserved water was transferred out of agriculture. However, such argument can be questionable from two perspectives. On one hand, agricultural water more than what was actually conserved through irrigation project rehabilitation may be taken away from farmers due to overestimation of water saving potential in the overall water rights transfer plan. On the other hand, the amount of water that can be abstracted under water abstraction rights (permits) in the Yellow River Basin was determined based on the average level of water availability in normal rainfall years. As part of water rights was transferred out from agriculture for as long as 25 years, the risk of reduced water availability in drought years had to be borne by farmers.

There may be wider impacts on communities beyond agriculture. The change in types of water use results in changes in both quantity and quality of return flows, which may lead to negative impacts on other users and the ecosystem at large. Unfortunately, these interests are far less of concern than economic benefits for some key stakeholders. Neither the provincial water department nor the river basin authority would check whether the water transfer may potentially damage water users, other stakeholders in community and the environment, although the ultimate goal of these pilot practices was to prevent the Yellow River from further deterioration.

The 2016 Provisional Measures on Administration of Water Rights Trading highlighted that water rights trading should benefit efficient utilisation and conservation of water resources, and at the same time, should not affect public interests and lawful rights and interests of stakeholders.[65] It further required, as part of the application for approval of water rights trading, the prospective transferor to submit information about the potential impacts of the transfer on public interests and other stakeholders' lawful rights and interests, as well as compensation strategies.[66] Yet, no detailed rules have been established regarding this important issue.

Another major challenge for China to develop an underpinning water rights administration system for water trading is how to design its water trading market within its institutional framework of hierarchic administration of water rights. While other overseas jurisdictions generally have one governmental authority in charge of water rights administration, China divides the responsibility into a hierarchy of water authorities at multiple levels, from river basin authorities down to county water bureaus. Such institutional arrangement makes verification of water rights

[65] 2016 Provisional Measures on Administration of Water Rights Trading, art. 5.

[66] Ibid, art.14.

practically difficult and adds transaction costs to market participants. Indeed, the fragmented water management based on administrative units (as low as county levels) presented as a key hindering factor to a wider water rights trading market in the Yellow River Basin. In order for the market to grow, China needs to work out a way to overcome this administrative barrier. If water rights administration was undertaken on a basin approach or by a uniform agency, the regulation of the water trading market would be a much easier task for China. However, this does not seem politically feasible as it requires substantial reform of China's political and administrative system. The recognition of river basins as water management units in China reflected such political challenge. As elaborated in Chap. 3, the 2002 Water Law established a combined water resources management approach based on administrative units and river basins.[67] Despite the importance of the integrated river basin management, the law failed to completely abandon the traditional fragmented water management based on administrative units. Following the vertical system of water resources administration called the Competent Department of Water Administration (CDWA), China's water rights administration will likely take a similar approach.

Based on the above analyses, this chapter further explores how to reform the current water rights administration system in China to facilitate a wider and more dynamic water trading market without a need to substantially change its fundamental institutional arrangements. Before that, the next section provides a review on overseas practices in water rights administration, for China to learn from the international experiences to build its legal, administrative, and regulatory infrastructure underpinning the growth of water trading markets.

7.3 Reflections on Experiences from Overseas

7.3.1 Centralised Water Rights Administration

7.3.1.1 State Ownership or Control: Legal Basis for Centralised Water Rights Administration

Roman law has directly influenced water laws in both common law and civil law jurisdictions, considered *res publici* (things owned by the public) and rivers (perennial ones other than the freshets) subject to authority of a centralised administration.[68] The right to divert water from public rivers was available to the public, subject to administration by the state.[69] Flowing water itself in rivers was viewed as *res communes* (a thing owned by all), which could not be the subject of ownership by

[67] For details, see Sect. 3.2.1.1.

[68] Anthony Scoot and Georgina Coustalin, "The Evolution of Water Rights," *Natural Resources Journal* 35, no. 4 (1995): 835. In Roman law, freshets which flowed only in periods of high rainfall or thaw were considered privately owned by the persons whose land they crossed.

[69] Ibid.

its physical nature but by the usufructuary rights.[70] This natural law was applied by common law judges in the eighteenth century to recognise the prior appropriative right since Sir William Blackstone went back to Roman law to seek theoretical foundation for appropriation.[71] As he observed,

> But after all there are some few things which notwithstanding the general introduction and continuance of property must still unavoidably remain in common, being such wherein nothing but a usufructuary property is capable of being had and they still belong to the first occupant during the time he holds possession of them and no longer. Such (among others) are the elements of light, air and water which a man may occupy by means of his...mill...[72]

In spite of the recognition of the private usufructuary rights in common law, ownership of water resources remained inexplicit. As noted in Chap. 6, the development of water rights has demonstrated an international trend in which statutory systems are established to regulate and govern water rights.[73] Water legislation not only formally recognises common law water rights regimes, but also legitimately accommodates public interests in water resources by bringing the resource under the ownership or control by the state. As the 2006 FAO law research commented:

> the common law has not generally recognised the concept of ownership over flowing water resources even by the state, water legislation in common law jurisdictions has tended to declare a superior state control right over water.[74]

Even in Chile, where free water markets are favoured to an unprecedented degree with restricted power for the state to regulate water rights, the 1981 Water Code defined water resources as national property for public use.[75] As public property, waters are owned by the nation as a whole and the use thereof belongs to all of its inhabitants,[76] based on which the state retains the power to administer water rights.

Similarly in other jurisdictions, individuals can own a right to use water (usufruct) but cannot own the water resource itself over which the state retains ownership or control. For example, in 19 US jurisdictions where an appropriation system or a hybrid system of riparian and appropriation is in place, state control over water resources is established through constitutional or statutory provision as follows[77]:

- Water "belong(s) to the public" (Arizona, Nevada, New Mexico, Oregon)
- Water is "property of the state" (Idaho, Montana, North Dakota, Texas, Wyoming)

---

[70] Ibid, 836.

[71] Ibid, 855.

[72] William Blackstone, *Commentaries on the Laws of England Book* 14 (1789), quoted from Ibid, 855.

[73] For further details about statute as an origin of water rights, see Sect. 6.1.4.

[74] Stephen Hodgson, *Modern Water Rights: Theory and Practices* (Rome: Food and Agriculture Organization of the United Nations, 2006), 37.

[75] Carl J. Bauer, *Against the Current: Privatization, Water Markets, and the State in Chile* (Boston & London: Kluwer Academic, 1998), 34.

[76] Ibid.

[77] Getches, *Water Law in a Nutshell*, 85.

- Water is "property of the people of the state" (California, Colorado, South Dakota)
- Water is "property of the public" (Nebraska, Utah)

It should be noted that groundwater in the US jurisdictions is subject to a range of rules and doctrines, different from and independent of those pertaining to surface waters. Groundwater rights may be based on overlying land ownership or established uses.[78] Despite the private property notions, state control over groundwater is not inhibited in most US jurisdictions.[79] Administrative permit systems are applied to bring groundwater within state management in order to protect competing users and allocate resources for the public interest.[80]

In the Australian jurisdictions, control over water is also vested in state governments.[81] Rather than explicitly expressed in ownership terms, state control over water resources tends to be recognised as statutory water rights of the state by water statutes in Australian jurisdictions. For instance, New South Wales (NSW) defined "state's water rights" in Section 392 in its Water Management Act 2000 as follows:

> (1) For the purpose of this Act, the rights to the control, use and flow of:
> all water in rivers, lakes and aquifers, and
> all water conserved by any works that are under the control or management of the Minister, and
> all water occurring naturally on or below the surface of the ground,
> are the 'State's water rights'…[82]

This provision of state's water rights was followed by "abolition of common law riparian rights" in Section 393:

> Any right that the owner of riparian land would, but for this section, have at common law with respect to the flow of any river, estuary or lake through or past the land, or to the taking or using of water from any such river, estuary or lake, is hereby abolished.[83]

The conferral upon the state of its statutory water rights brings the taking, use and the carrying out of prescribed activities into an administrative licensing system.[84] Without claiming ownership, the state has placed its water resources under control.

Getches, the author of *Water Law in a Nutshell* (third edition) argued that the authority of water rights administration is rooted from sovereignty rather than property, though the state's interest is typically expressed in ownership terms by statutes

---

[78] Ibid, 247.

[79] Ibid, 253.

[80] Ibid, 253–254.

[81] Australia Productivity Commission. *Water Rights Arrangements in Australia and Overseas*, 49.

[82] 2000 Water Management Act of New South Wales, http://www.austlii.edu.au/au/legis/nsw/consol_act/wma2000166/ (accessed May 22, 2017), Section 392.

[83] Ibid, Section 393.

[84] Despite the abolition of common law riparian rights, basic landholder rights are vested by statute and allow water to be taken for stock and domestic purposes without approval. See Australia Productivity Commission, *Water Rights Arrangements in Australia and Overseas*, 70.

or constitutions.[85]As he pointed out, "governments use their policy power to regulate uses of water and to conserve and allocate the resources in the interest of the public".[86] Nevertheless, whether it is ownership or sovereignty, state control over water resources has to be exercised through institutional arrangements for water management, including administration of water rights.

7.3.1.2 The Administrative Approach to Water Rights Administration

With an administrative approach to water rights administration, an administrative agency is vested the authority to allocate and administer water rights. Permit procedures are created to regulate existing water rights and to issue new water rights. In some jurisdictions, water resource management tends to be more centralised with a central government department taking the responsibility. In Chile, for example, the government water rights agency is the General Water Directorate (*Dirección general de Aguas*, DGA) responsible for granting requests for new water rights.[87] In each of the Australian states, a minister is assisted by a government department to take the water management responsibility.[88] In NSW, for example, the Department of Primary Industries DPI Water (DPI Water) is responsible for water licensing.[89] Among other Australian jurisdictions, Victoria is an exception with a decentralised administration of water rights. While its Minister for Environment, Climate Change and Water is in charge of administering bulk entitlement orders with Department of Environment and Primary Industries, other water rights are administered by rural water authorities.[90]

Every western state in the US has vested the authority of administering water rights in an administrative agency, with a range of names such as the State Engineer, the Department of Natural Resources, or Water Resources Control Board.[91] In California, the State Water Resources Control Board is the principal regulator of surface water use, in which its Division of Water Rights administers California's

---

[85] Getches, *Water Law in a Nutshell*, 85,139.

[86] Ibid, 82.

[87] Bauer, *Against the Current*, 35. In addition to the DGA, local civil courts in Chile take the responsibility for recognizing traditional water rights.

[88] Australia Productivity Commission, *Water Rights Arrangements in Australia and Overseas*, 75.

[89] New South Wales Department of Primary Industries Water, *Water Licensing*, 2017, http://www.water.nsw.gov.au/water-licensing (accessed March 2, 2017). Based on the passage of the Water NSW Amendment (Staff Transfers) Bill through Parliament in 2016, a number of functions related to the delivery of water services in New South Wales will be transferred from DPI Water to WaterNSW from July 1, 2016.

[90] Australia Productivity Commission, *Water Rights Arrangements in Australia and Overseas*, 184.

[91] Getches, *Water Law in a Nutshell*, 138, 141. Colorado Division of Water Resources, also known as the State Engineer's Office, administers water rights. However, a system of water courts is in place to govern water law, to which stakeholders apply for a water right and oppose an application for a new water right. See Deserai A. Grow, "Policy Punctuations in Colorado Water Law: The Breakdown of a Monopoly," *Review of Policy Research* 27, no. 2 (2010):155.

water right system.[92] It is responsible for issuing appropriate permits, and approving transfers of surface water appropriated pursuant to a permit or licence issued by it, where the transfer led to changes in use such as the point of diversion, place of use, or purpose of use.[93]

7.3.1.3 The Judicial Approach to Water Rights Administration

Unlike other states in the US where the prior appropriation doctrine is followed, Colorado takes a judicial approach to water rights administration.[94]As the first US state to provide special court proceedings for water rights controversies, Colorado vested jurisdiction over all water rights adjudications in the district court of each county before 1969.[95] The 1969 Water Rights Determination and Administration Act centralised the jurisdiction to seven water courts by dividing the state into seven water divisions that correspond to the seven major drainages.[96] Existing water rights, new water rights, water rights transfers, and changes of use are approved by water referees or water judges, and evidenced by court decrees. In addition to water courts, the State Engineer plays a role in assisting the courts to administer water rights. The following figure (Fig. 7.3) illustrates a general procedure of Colorado's water rights adjudication.[97]

To summarise, it can be observed that water rights administration tends to be centralised in most jurisdictions, with a few exceptions. Among jurisdictions where an administration approach is employed, Victoria in Australia allows water rights other than bulk entitlements to be administered by rural water authorities who issue water rights to water users within their irrigation districts. Indeed, water rights in this context are an instrument similar to water supply contracts or company shares provided by water supply organisations or water user associations in other jurisdictions. Generally speaking, administrative permitting systems to administer water

[92] California State Water Resources Control Board, *Water Rights Applications: Permitting and Licensing Program*, 2017, http://www.swrcb.ca.gov/waterrights/water_issues/programs/applications/ (accessed March 2, 2017).

[93] Brian E. Gray, "The Market and the Community: Lessons from California's Drought Water Bank," *Hasting West-Northwest Journal of environmental Law & Policy* 14 (2008): 64.

[94] Grow, "Policy Punctuations in Colorado Water Law," 155. It is important to note that in Colorado, the right to withdraw designated ground water is handled by the Colorado Ground Water Commission via the Office of the State Engineer. There is also a well permit system in place, again handled by the Office of the State Engineer. See Colorado Division of Water Resources, *Guide to Colorado Well Permits, Water Rights, and Water Administration*, 2012, http://water.state.co.us/DWRIPub/Documents/wellpermitguide.pdf (accessed May 22, 2017).

[95] Getches, *Water Law in a Nutshell*, 152.

[96] Ibid, 153.

[97] The procedure is summarised from description of the Colorado system in Getches, *Water Law in a Nutshell*, 153–154.

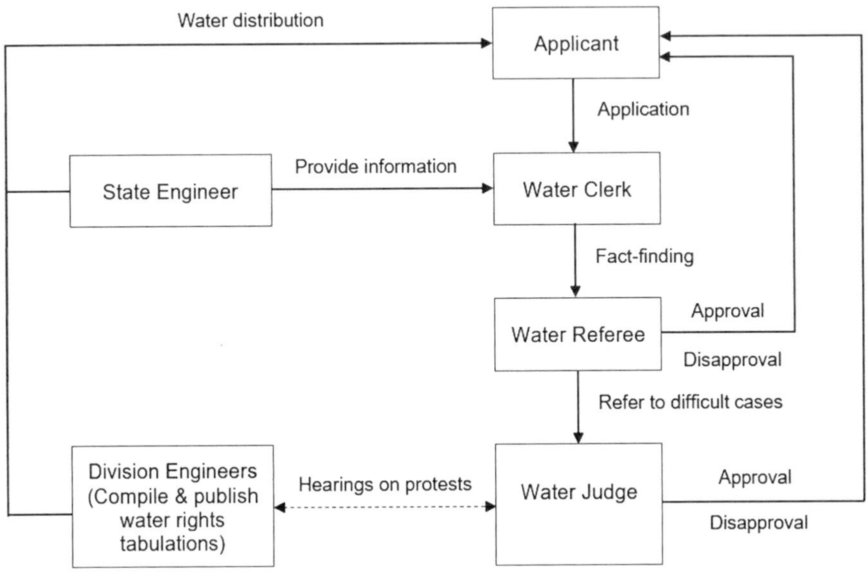

Fig. 7.3 Procedure of water rights adjudication in Colorado

rights tend to be centralised, in which case the authority tends to lie within a central government agency.

In Colorado, although it does not have a permit system for water rights, seven water courts share similar responsibilities to administrative permitting agencies in other jurisdictions. While taking a more decentralised approach, Colorado's judicial system for water rights adjudication experienced an evolution of centralisation. Its 1969 water rights statute vested the function of water rights administration that used to be shared by county district courts into seven drainage-based water courts.

7.3.2 Sophisticated Titling/Registration Systems

Water rights have been usually viewed as a real property right. Unlike chattels whose title is evidenced by possession, water rights tend to be registered like other real estate properties as evidence of the right. By providing certainty to the legal existence of a property right, a titling/registration system plays a critical role in underpinning the operation of a property rights regime as well as markets of these property rights.[98] Registers of water rights need to be maintained to record issuance,

[98] ACIL Tasman and Freehills, *An Effective System of Defining Water Property Titles*, Research Report to the Australian Government Department of Agriculture, Fisheries and Forestry, and Land & Water Australia, Canberra: Land & Water Australia, 2004, 21. Also see Michael Woolston, "Registration of Water Titles: Key Issues in Developing Systems to Underpin Market Development," in *The Evolution of Markets for Water: Theory and Practice in Australia*, ed. Jeff Bennett (Cheltenham, UK; Northampton, MA, Edward Elgar, 2005), 80.

modification, and transfer of water rights, as an essential element of water rights administration. By assuring the security of water rights, titling/registration systems can be crucial to facilitate water rights trading. Efficient market operation requires not only security of water rights but also reasonable transaction costs to encourage trading. As a result, how a water rights trading market operates is closely related to accessibility of the water rights titling/registration systems for market players to verify title.

7.3.2.1 Organisations for Water Rights Titling/Registration

Water rights registers are typically maintained in governmental agencies responsible for water rights administration. Some jurisdictions, however, may also have a separate registry for water rights. In Queensland, for example, while its water administration agency (Department of Natural Resources and Mines) maintains a database of granted water allocations and interim licences, water allocations are recorded on the Water Allocations Register, which is a module of the Queensland Resource Registry.[99] In addition to the separated approach to water rights registry, there also has been institutional separation to handle different types of water rights trades. The register of the Department of Natural Resources and Mines tracks temporary trades while the Queensland Resource Registry only deals with permanent trades and other defined interests.[100]

Accessibility of water rights registers is an important transaction cost factor for water trading. Jurisdictions have also been making efforts to improve public access to information about water rights registers. In Australian jurisdictions, online searching of water rights registers is available to the public so that verification of the title is speedy and less costly. In some jurisdictions, this process may be more complicated and time consuming, such as in Colorado. With a few exceptions, water rights in Colorado are generally transferred by a deed recorded in the clerk and recorder's office of the county where the water right is located.[101] The State Engineer's Office does main a database of water rights including those decreed through the water courts.[102] However, although this central database is accessible online which can be searched by location,[103] a water rights trade in Colorado most often involves a water attorney to perform "due diligence" to assess whether the

---

[99] Australia Productivity Commission, *Water Rights Arrangements in Australia and Overseas*, 185, 204.

[100] Woolston, "Registration of Water Titles," 86.

[101] P. Andrew Jones and Tom Cech, *Colorado Water Law for Non-Lawyers* (Boulder: University Press of Colorado, 2009), 227.

[102] Colorado Division of Water Resources, *Guide to Colorado Well Permits, Water Rights, and Water Administration*, 11.

[103] Ibid.

water right concerned in the trade has good title.[104] The water attorney will not only have to obtain documents from the local recording office to review the recorded deeds, but also conduct a field visit to learn more about the history of the right.[105] In many cases, a water engineer may also be engaged in the "due diligence" process to provide technical information.[106]

7.3.2.2 Types of Water Rights Titling/Registration Systems

In Australia, there are researchers who advocate the application of the Torrens title system in the registration of land ownership to register water rights, which falls into the category of a "registration" system, or more technically speaking, "registers of rights".[107] Under Torrens title, details of a water right are recorded in a central register including rights and liabilities of the right, changes made to the right, and financial interests in the right.[108] Subject to certain exceptions, the registered proprietor obtains an indefeasible title to the water right, which means that their title "cannot be affected or defeated by any existing estate or interests, other than registered interests noted in the register".[109] As the report by TCIL Tasman and Freehills commented, with indefeasibility as its fundamental principle,

> A Torrens-based system is a system of 'title by registration' rather than a system of 'registration of title'. Dealings take effect upon registration, rather than upon execution of the documentation relating to the dealing.[110]

Accordingly, Torrens systems allow persons dealing with a water right to rely on the register as proof of the quality of title without need to search the chain of title for the right, which contributes to lowering transaction costs.[111] Where the integrity and accuracy of the titling system is guaranteed by government, the public and investors in particular can be more confident in the security of the right, and thereby help facilitate trading markets[112] In Australian jurisdictions, none has adopted indefeasibility in respect of its water rights registers as that adopted in the Torrens land title system.[113] Some proposed that governments should be prepared to do so as they

---

[104] Jones and Cech, *Colorado Water Law for Non-Lawyers*, 238.

[105] Ibid, 239.

[106] Ibid, 241.

[107] ACIL Tasman and Freehills, *An Effective System of Defining Water Property Titles*, 22. Woolston, "Registration of Water Titles," 81.

[108] Australia Productivity Commission, *Water Rights Arrangements in Australia and Overseas*, 103.

[109] ACIL Tasman and Freehills, *An Effective System of Defining Water Property Titles*, 22. Woolston, "Registration of Water Titles," 81.

[110] Ibid, 41.

[111] Ibid, 43.

[112] Ibid, 44.

[113] Woolston, "Registration of Water Titles," 88.

have committed to the refinement of water property rights and the promotion of water trading markets in the context of its current water reform.[114]

In contrast with "registers of rights" under the Torrens system, a "recording system", more known as "registers of deeds", does not guarantee legal ownership of water rights. Most US jurisdictions, including California and Colorado, as well as some Australian jurisdictions such as Victoria have used this conventional registration system. Without providing indefeasible titles, it requires a person who intends to deal with the right to search written records of previous dealings in relation to the right for verification of title.

7.3.3 Accessible Trading Platforms and Expanded Trading Range

7.3.3.1 Various Forms of Water Markets

As noted in Chap. 6, temporary transfers represent a main type of water rights transaction in many jurisdictions where water rights trading markets are operated.[115] Temporary trading plays an important role in managing variability and uncertainty of water availability thanks to their flexibility. For this reason, what matters most for temporary transfers is that water can be available exactly when it is needed. In other words, this type of transaction requires a simplified and smooth process to pursue speed and efficiency. To achieve this, jurisdictions have developed different forms of market designs, such as water banks, online trading platforms, and the use of water brokers.

Water banks have been an innovative and successful design of California and have been increasingly used in wider regions of the US. As early as the severe drought in 1976–1977, a water bank was set up by the Bureau of Reclamation (BoR) to facilitate temporary water trading among public water entities.[116] In 1991, when California entered its fourth consecutive drought year, a water bank was again created through which the Department of Water Resources could purchase water for resale to areas of California suffering from supply shortages.[117] Unlike the BoR water bank, this state-wide water bank allowed both public and private water users, especially farmers, to buy and sell water on a 1-year basis.[118] Despite criticism from

[114]ACIL Tasman and Freehills, *An Effective System of Defining Water Property Titles*, 44.

[115]For further details, see Sect. 6.2.4.

[116]Charles W. Howe, "Increasing Efficiency in Water Markets: Examples from the Western United States," in *Water Marketing – the Next Generation*, ed. Terry L. Anderson and Peter J. Hill (Lanham, Boulder, New York, London: Rowman & Littlefield Publishers, Inc., 1997), 86.

[117]Brent M. Haddad, *Rivers of Gold, Designing Markets to Allocate Water in California* (Washington, D.C.: Island Press, 2000), 51; Gray, "The Market and the Community," 48.

[118]Howe, "Increasing Efficiency in Water Markets," 87.

those who believed that the water bank had blocked the opportunity for a true market price to be realised,[119] the arrangement was widely acknowledged as a success.[120] One remarkable achievement of the water bank was reallocating water between users in a less costly and time consuming manner and thereby helping the state through its emergency drought. As one author commented,

> It undoubtedly increased the feasibility of short-term water transfers, and it paved the way for an unprecedented number of north-south transfers in a short period of time. It swept aside technical (conveyance) barriers to north-south transfers and accelerated the transfer-approval process.[121]

Trading platforms are another key element of water markets. Australia's water reform has seen refinements in trading platforms of water markets, including the introduction of electronic exchanges, the emergence of water market intermediaries (brokers), and more accessible and publicly available information, to help reduce the cost of bringing buyers and sellers together and enabling them to make well-informed decisions.[122] In the late 1990s and early 2000s, governments, water businesses, and private brokers in the Murray Darling Basin started introducing electronic centralised exchanges as trading platforms for water markets, predominantly facilitated by brokers.[123] Major exchanges currently operating as water trading platforms in Australia include Water Exchange, Murrumbidgee Water Exchange, Murray Irrigation Water Exchange, and Waterfind.[124] These water trading platforms "match buyers and sellers, either through an automated process or a bulletin board",[125] many of which also provide brokerage services.[126] Often conducting trades through exchanges, water brokers, many of whom are estate agents, emerged to help irrigators through the water trading process. Although there is lack of specific regulation, water market intermediaries are subject to general legislative measures, which offer protection against broker misconduct.[127] In 2007, the Australian Water Brokers Association was established and developed a code of ethics and standards for its members.[128] It is estimated that approximately 80% of water trades are facilitated by intermediaries (55% occurring on water exchanges and 25% through

[119] Gray, "The Market and the Community," 55.

[120] Ibid. Also Haddad, *Rivers of Gold*, 55.

[121] Gray, "The Market and the Community," 56.

[122] Australia National Water Commission (NWC), *Water Markets in Australia: A Short History*, 2011, Canberra: National Water Commission, 11.

[123] Ibid, 68.

[124] Australia National Water Commission (NWC), *Australia Water Markets Report 2012–13*, 2013, Canberra: National Water Commission, 25.

[125] Australian Competition and Consumer Commission (ACCC), *Water Market Intermediaries – Industry Developments and Practices*, 2010, Canberra: Australian Competition and Consumer Commission, 12.

[126] Ibid.

[127] NWC, *Australia Water Markets Report 2012–13*, 23.

[128] Ibid.

brokers).[129] These water market intermediaries play an important role in the creation and operation of water markets, as they reduce transaction costs, encourage a deeper market, and provide price and other market information.[130]

7.3.3.2 Inter-jurisdictional Transfers

Diversity of water property rights regimes and water trading rules across jurisdictions have formed institutional barriers that limit and/or prevent interstate water transfers. To maximise the functions of water rights trading in water reallocation, some jurisdictions attempted to expand the trade in water beyond the boundaries of states. Australia, for example, has been making substantial efforts to broaden and expand the geographical scope of its water markets. The Murray-Darling Basin (MDB), especially the southern MDB, is where the majority of Australia's water market activity takes place. Collaborative inter-jurisdictional efforts are made to facilitate the growth of water trading across connected water systems and state borders in the MDB, including the establishment of a consistent regulatory environment across the MDB.

In 1992, the Commonwealth, NSW, Victorian, and South Australian governments signed a Murray-Darling Basin Agreement, which incorporated arrangements for promoting interstate water trading.[131] The Murray-Darling Basin Commission (MDBC) is responsible for managing the allocation and delivery of water to the three River Murray states, advising on water, land, and other environmental issues throughout the river basin, as well as playing a significant role in facilitating interstate water trading in conjunction with the states.[132] In 1997, the Murray-Darling Basin states and the Commonwealth Government capped the level of water diversions from the basin at 1993–1994 levels of development.[133] A cap on water extractions in the MDB was established for each jurisdiction, serving as a major impetus for water trading as a water reallocation mechanism with an objective to ensure environmental sustainability.[134]

The Interstate Entitlement Trading Pilot Project was commenced in 1998, enabling both entitlement and allocation trading across state boundaries of NSW, Victoria, and South Australia.[135] The 2004 National Water Initiative (NWI) further endorsed interstate entitlement trading, aiming to develop a nationally consistent and compatible water market.[136] Interstate water trading has expanded to the entire southern connected MDB. Tagging arrangements were introduced in 2007, through

[129] ACCC, *Water Market Intermediaries*, 12.

[130] NWC, *Water Markets in Australia*, 68.

[131] Ibid, 48.

[132] Ibid.

[133] Ibid, 53.

[134] Ibid, 6.

[135] Ibid, 60.

[136] Ibid, 50.

which the entitlement is tagged in its state of origin for use in the state of destination.[137] While tagged entitlement trading was rare in practice, strong market performance is achieved in interstate water allocation trading.[138] Moreover, the Water Act 2007 provides the legal framework for the Australian Government and its agencies to play a role in establishing, monitoring and enforcing compliance with water trading, water market, and water charge rules.[139] More recently, the 2012 Murray-Darling Basin Plan has provided a coordinated approach to water use across the basin's four states (Queensland, NSW, Victoria, and South Australia) and the Australian Capital Territory, which incorporates rules for the trading in a range of tradable water rights (e.g. water access rights, irrigation rights, and water delivery rights).[140]

7.3.3.3 Trading in Water Rights for the Environment

As the values of water for non-consumptive purposes such as recreation, ecosystems, and the environment are increasingly appreciated, people have come to recognise instream or environmental uses of water besides abstractive uses (e.g. domestic, industrial, and agricultural uses). Jurisdictions have implemented mechanisms, such as minimum stream-flow requirements, or issuance of new water rights for instream flows, to protect the environmental needs for water. However, the effectiveness of these measures have remained limited as the new instream rights are acquired long after the earlier appropriative rights and thereby are relatively junior in priority. As a policy response to this issue, trading in water rights is increasingly considered an alternative solution to account for these critical values of water resources.

In the western US, where water rights are based on appropriation activities, the concept of beneficial use has been interpreted to include instream uses for recreation, aesthetics, and maintenance of ecosystems. Colorado's 1969 Water Right and Determination Act expanded the range of beneficial uses by including impoundment for recreation, such as fishing and wildlife and maintenance of minimum flows necessary for the reasonable preservation of the environment as viable forms of appropriation to establish water rights that can be held by the Colorado Water Conservation Board (CWCB).[141] The 2001 amendments of the Act further established appropriations for recreational in-channel diversions (RICDs) to allow local governments including counties, municipalities, cities, water districts, water and sanitation districts, water conservation districts, or water conservancy districts to become instream right holders.[142] In California, acquisition programmes have also

[137] NWC, *Australia Water Markets Report 2012–13*, 6.

[138] Ibid.

[139] NWC, *Water Markets in Australia,* 87.

[140] Ibid, 94.

[141] Rebecca Abeln, "Instream Flows, Recreation as Beneficial Use, and the Public Interest in Colorado Water Law," *University of Denver Water Law Review* 8 (2005): 522.

[142] Ibid, 524.

been established for the conservation or restoration of threatened or endangered species. Different to Colorado's requirements on instream right holders being the state, California allows any person or entity to own or dedicate an existing water right to instream beneficial use for fish and wildlife and recreation benefit.[143] By allowing market forces to drive instream flow transfers, late stream uses can obtain a senior water right whose original priority date can be retained.

Besides acquisition programmes of governmental agencies (both federal and state), private organisations are also involved in market transfers to protect and restore instream flows in states where private parties are allowed to hold instream rights. One prominent example is the use of Water Trusts in Oregon. The 1987 Instream Water Rights Act of Oregon recognised instream uses as beneficial and allowed the appropriation and transfer of instream water rights.[144] Based on this provision, private organisations are able to acquire water rights through sale, lease, and donation, and transfer such rights to the Water Resources Department that holds the rights on behalf of the people of the State of Oregon.[145] As a private, non-profit organisation, the Oregon Water Trust was established in 1993 based on the land trust model to engage in market acquisitions of instream rights.[146]

In Australia, based on the "cap and trade" approach to water markets, the Commonwealth Government has made significant financial commitments (AUD $3.1 billion over 10 years) to buy back water entitlements from irrigators in the MDB for the environment.[147] The Commonwealth Government also acquires water entitlements through its Sustainable Rural Water Use and Infrastructure Programme involving a commitment of AUD$5.8 billion over 10 years to subsidise investment in irrigation infrastructure.[148] The Commonwealth Environmental Water Holder (CEWH) is responsible for managing all entitlements purchased by the Australian Government.[149] Government agencies have been playing an increasingly important role in water markets and the CEWH, for example, regularly transfers allocations between regions.[150] In 2012–13, large volumes of interstate water allocation transfers were for the delivery of environmental water, especially from Victoria to South

---

[143] Charlton H. Bonham, "Perspectives from the Field: A Review of Western Instream Flow Issues and Recommendations for a New Water Future," *Environmental Law* 36 (2006):1223.

[144] Mary Ann King, "Getting Our Feet Wet: An Introduction to Water Trusts," *the Harvard Environmental Law Review* 28 (2004): 505.

[145] Ibid, 506.

[146] Janet C. Neuman, "the Good, the Bad, and the Ugly: the First Ten Years of the Oregon Water Trust," *Nebraska Law Review* 83 (2004): 436.

[147] NWC, *Water Markets in Australia,* 78.

[148] Ibid, 80.

[149] Ibid.

[150] Australian Bureau of Agricultural and Resource Economics and Sciences (ABARES), *Australian Water Markets Report 2014–15*, 2016, Canberra: Department of Agriculture and Water Resources, 51.

Australia.[151] The emergence of environmental allocation transfers has further exaggerated growth in allocation trade volumes in more recent years, as in 2014–15, environmental transfers in the southern MDB contributed to 38% of all allocations traded in Australia.[152]

7.3.4 *Increasingly Accommodated Third-Party Effects*

7.3.4.1 Transfer Approval and Restrictions to Protect Third-Party Interests

As a result of the integral and fugitive nature of water, water users who share the same water resource are interdependent with each other, which means that where and how much a water user has used for their purpose, would have impacts on others within the same water cycle. The impacts on third parties would therefore easily take place where a water rights holder seeks a transfer that involves changes in the amount, the purpose, and the location of their existing use. "Third parties" here not only refer to other water right holders, but also include wider groups interested in water transfers such as environmentalists and other communities including indigenous people that value water for its economic, social, and cultural good. Focusing on their bilateral interests, however, the transfer parties can be expected to "pay scant if any attention to the interests of the third parties who will bear the external costs of the transactions".[153]

Since the market transfer itself can barely account for these effects on third parties, it is necessary for government to provide institutions and take measures to ensure that markets operate efficiently in water reallocation and with minimised negative impacts for interest groups apart from the two parties in the transfer. As the concept of "third parties" is used in its wider scope to include public values of water, the protection of their interests remains government's responsibility. Therefore, jurisdictions tend to impose various restrictions on water rights trading to accommodate the interests of third parties. Jurisdictions normally require water rights holders who seek a water transfer that will potentially result in changes in water use to apply to permitting authorities for an approval.

In the Australian jurisdictions, water rights transfers are generally assessed by water rights administrators against a range of administrative criteria. Generally developed from overarching legislation, regulation, guidelines and resource plans, such criteria assist the administrators to determine whether the proposed transfer will cause harm to other rights holders, water uses, the environment, or the community more widely.[154] Agricultural and riparian land, rivers and streams are often

---

[151] NWC, *Australia Water Markets Report 2012–13*, 6.

[152] ABARES, *Australian Water Markets Report 2014–15*, 2.

[153] Gray, "The Market and the Community," 96.

[154] Australia Productivity Commission, *Water Rights Arrangements in Australia and Overseas*, 196.

divided into zones with common features so that the administrative processes can be streamlined.[155] State governments have introduced approval processes for water rights trade. Trades in water access entitlements have to follow more stringent approval processes than those for water allocations as the entitlement trades have longer-term implications for third parties.[156] Restrictions on trade of entitlements out of irrigation regions (4% limit) are also in place to address the concerns about the stranded assets and social impacts on the irrigation region and broader community.[157]

Compared to the Australian jurisdictions whose trading rules have comprehensively defined or applied application assessment matters to examine third-party effects, the US states impose common law and statutory restrictions on water transfers with fewer precise standards to guide their decisions. Three types of restrictions, the no injury rule, the area of origin protection, and the public trust doctrine, are further observed in the following text.

7.3.4.2 The No Injury Rule

All western states of the US adopt no injury rules to protect return flows of downstream users.[158] Transfer applications are subject to an examination of the no injury rule, which focuses on protection of return flows of downstream users to whom water returned back to the stream by the appropriative right holder is available for appropriation.[159] In a typical case, permitting authorities (either an administrative agency or a water court) file a transfer application, and notice of the proposal is published. Other right holders can object to the proposed transfer by claiming injury in that it will result in an increase in water consumption by the transferee, changes in stream flow, changes in the time, frequency or length of diversion, or changes in water quality.[160]

Protection of the broader range of social and environmental values leads to broadening of the "no injury" concept. In some cases, injury claims of third parties

[155] Ibid, 198.

[156] NWC, *Water Markets in Australia*, 61.

[157] Ibid, 63.

[158] Barton H. Thompson, "Institutional Perspectives on Water Policy and Markets," *California Law Review* 81 (1993):703.

[159] Ibid. In California as an example, because the state does not have jurisdiction over most groundwater-related transfers, the protection of third parties by no injury laws do not fully extend to groundwater users. To correct this omission, rural counties have their own local ordinances to restrict groundwater exports. See generally Ellen Hanak and Caitlin Dyckman, "Counties Wresting Control: Local Responses to California's Statewide Water Market," *University of Denver Water Law Review* 6 (2003): 490–518.

[160] Megan Hennessy, "Colorado River Water Rights: Property Rights in Transition," *University of Chicago Law Review* 71 (2004):1670.

can also stand on environmental or public interest grounds.[161] The burden of proof typically rests on the parties seeking to transfer water rights, which makes the rule serve the important function of protecting third parties but in the meantime placing restraints on water transfers.[162] Some observers argued that the no injury rule appears to be one of the critical obstacles in transferring water rights, which undermines the effectiveness of water markets.[163]

7.3.4.3 The Area of Origin Protection

In addition to the no injury rule that protects existing legal users of water, restrictions on interbasin transfers are also imposed in most western states of the US to protect the interests of areas of origin whose local economy, ecology, lifestyle, and potential for future growth are likely to be negatively affected by water diversion out of the area.[164] The area of origin protection usually provides for either the right of the area of origin to "recapture" the water at some time in the future, or to be compensated for its continued loss, which can be divided into four categories[165]:

- Prohibitions seek to keep the water in the watershed to preserve and protect the economic, environmental, cultural, and other in situ values within the watershed (such as Arizona, Montana).[166]
- Restrictions are imposed on the interbasin transfer process to ensure that public and local interests are considered in the transfer decision (such as Kansas, Idaho, Oregon, Texas).[167]
- Recapture provisions allow the water export and grant the originating basin a priority right to the water based on which the basin of origin can recapture the exported water when it needs it for development (such as California).[168]
- Only surplus water can be exported out of the basin and the originating basin retains the amount of water needed to meet their needs for economic development and environmental protection (such as California, New Mexico, Texas).[169]

[161] Ibid.

[162] Ibid,1671.

[163] For example, see Caitlin S. Dyckman, "A Dynastic Disruption: The Use Efficiency and Conservation Legacy of the Governor's Commission to Review California Water Rights Law Recommendations," *McGeorge Law Review* 36 (2005):197.

[164] Ronald A. Kaiser, "Texas Water Marketing in the Next Millennium: A Conceptual and Legal Analysis," *Texas Tech Law Review* 27 (1996):215.

[165] Ibid, 216.

[166] Ibid.

[167] Ibid, 217.

[168] Ibid, 218.

[169] Ibid.

7.3.4.4 The Public Trust Doctrine

Flexibly applied in the environmental and natural resources contexts, the public trust doctrine in the US jurisdictions has also been a means to protect third-party interests involved in water allocation and reallocation. Originally, the doctrine held the state ownership of navigable water bodies and related resources in trust for the benefit of the people of the state.[170] It began to be loosened from its historical connection with navigation and waterways and applied as a general device for incorporating community values in natural resources since Sax's 1970 article "The Public Trust Doctrine in Natural Resource Law: Effective Judicial Intervention" called for expanded applications of the doctrine.[171] In 1983, the California Supreme Court revolutionarily expanded the public trust doctrine to encompass the appropriation of water from surface watercourses in the case of the National Audubon Society vs. Superior Court.[172] To connect the trust with the appropriative rights doctrine, "the court reiterated both the state's continuing supervision over previously approved appropriation, and the state's ability to reconsider past allocation decisions which may be incorrect in light of current knowledge or inconsistent with current needs".[173] Since then, the public trust doctrine has been developed in both legislative and administrative contexts with respect to water rights transfer in California, although the doctrine's application to water diversion has little judicial development both in and outside California.[174] As the permitting agency, the SWRCB has addressed public trust matters in new applications to appropriate water, temporary water transfers, and existing appropriation.[175] According to the California Water Code Sections 1425 and 1435, the SWRCB has been expressly directed to determine whether proposed water transfers will have unreasonable effects on fish, wildlife, or other instream beneficial uses.[176] The doctrine allows not just traditional instream values but the impact of a water diversion on the entire ecosystem at the diversion point, such as air resources and wildlife habitat, to be considered in the process.[177]

---

[170] Generally see Joseph Sax, "The Public Trust Doctrine in Natural Resource Law: Effective Judicial Intervention," *Michigan Law Review* 68 (1970): 475–566; Carol M. Rose, "A Review of Professor Joseph Sax's Defense of the Environment through Academic Scholarship: Joseph Sax and the Idea of the Public Trust," *Ecology Law Quarterly* 25 (1998): 351–362; Cynthia L. Koehler, "Water Rights and the Public Trust Doctrine: Resolution of the Mono Lake Controversy," *Ecology Law Quarterly* 22 (1995): 541–589; George P. Smith and Michael W. Sweeney, "The Public Trust Doctrine and Natural Law: Emanations within A Penumbra," *Boston College Environmental Affairs Law Review* 33, no.2 (2006): 307–343.

[171] Generally see Sax, "The Public Trust Doctrine in Natural Resource Law"; Rose, "A Review of Professor Joseph Sax's Defense of the Environment through Academic Scholarship".

[172] Gregory S. Weber, "Articulating the Public Trust: Text, Near-Text and Context," *Arizona State Law Journal* 27 (1995):1160.

[173] Ibid,1162.

[174] Ibid,1171.

[175] Ibid,1173.

[176] Ibid,1172, 1178.

[177] Ibid,1229.

Remaining vague and evolving, the public trust doctrine has been expected by some observers to continue to increasingly play important roles in water allocation and reallocation, while others call for attention to necessary restraint on its expansion.[178] It reflects to some degree the dilemma in the matter of government and market as well as the matter of private property rights and public values.

7.4 Implications and Recommendations for China's Water Reform

7.4.1 Water Rights Administration: Decentralisation or Centralisation?

In comparison to the overseas jurisdictions observed previously, China seems to have created a unique style of administering water rights. It falls in the group of the administrative permitting approach, yet in a much more decentralised fashion. The water rights administration responsibility has been divided both horizontally and vertically based on the multi-level administrative system of governments.

At the top level, the MWR is the principal regulator of China's water use. Under the existing water regulations, however, the MWR delegates the powers and functions of water rights administration down to major river basin (or lake) authorities. At the local levels, provincial, prefecture, and county water departments also share responsibilities corresponding to their jurisdictions. It means that water rights in China are administered by seven river basin/lake authorities as well as the local WRBs/WABs in 34 provincial,[179] 333 prefecture, and 2862 county governments.[180] This decentralised approach to water administration might help distribute workload for water authorities and take effective control over local water uses. In terms of facilitating water rights trading, however, such decentralisation in water rights administration has been proved to be a critical impediment to the growth of water markets.

[178] Ibid,1248.

[179] 34 provincial governments include 4 municipalities directly under the State Council, 5 autonomous regions, 22 mainland provinces, Hong Kong Special Administrative Region, Macao Special Administrative Region, and Taiwan Province. However, provincial governments in the book exclude Hong Kong, Macao, and Taiwan.

[180] Data from the official website of the Department of District Division and Place Naming, China Ministry of Civil Affairs, http://www.xzqh.org.cn/ONEWS_zq.asp?id=1150 (accessed August 28, 2008).

7.4.1.1 Is Centralised Water Rights Administration at the Basin Level Possible?

As a unitary country, China has long been in a dilemma in the division of central-provincial authority. Instead of a formal institutional arrangement for such division, the relationship between central and provincial governments is thrashed out in case-by-case agreements, subject to renegotiation as conditions and needs change.[181] As a result, one can see alternating periods of decentralisation and recentralisation taking place in China. Its economic reform and market transition that began in the late 1970s significantly enhanced the role of local governments in economy as well as public administration. As a 2005 World Bank report commented: "With decentralisation, subnational governments are now at the heart of a range of investments, fiscal, and regulatory activities that affect both the pace and quality of economic growth".[182]

Not surprisingly, local governments have formed their own political and economic interests, independent of the central government and of other localities. One of the major side effects of such local interests is environmental degradation. In the case of water resources, it is not uncommon that upstream provinces over-abstract water to meet rapidly increasing needs of its local economy without considering the need of downstream provinces for water and the health of the water systems. Water pollution, particularly transboundary pollution is difficult to control as the ones suffering the consequences are usually the downstream areas other than the polluting province.

To address environmental problems, the central government recentralised authority in natural resources management by strengthening central control of relevant ministries under the State Council over their local counterparts. For example, although it is politically difficult, the MEP has been seeking direct management of its local EPBs, whose personnel is currently employed and financed by local governments. As a transitional approach, five regional supervision centres were set up to take responsibility for supervising transboundary environmental protection and coordinating transboundary pollution disputes,[183] but their roles remained limited due to the ambiguous legal status and the awkward hierarchic pitch much lower than provincial governments.

River basin authorities have faced a similar challenge. Water management based on river basins was recognised in legislation, and increasingly implemented in practice. Seven river basin/lake authorities for China's major water systems have been strengthened, though they still struggle to counterbalance the administrative power of provincial governments who are in a higher stand in China's administrative hierarchy.

---

[181] John Bryan Starr, *Understanding China* (London: Prfile, 1998), 71.

[182] World Bank, *East Asia Decentralizes: Making Local Government Work*, Washington DC.: World Bank, 2005, 2.

[183] Information from the official website of Ministry of Environmental Protection of China, http://www.zhb.gov.cn/hjjc/hjzf/djt/200612/t20061229_99016.htm (accessed August 28, 2008).

To overcome these institutional constraints, two reform proposals were raised that either river basin authorities be upgraded to a level directly under the State Council instead of as the MWR's agencies, or that water resources committees for water rights management be set up at the national and river basin levels.[184] Requiring substantial changes in China's political and administration system, both proposals seem unrealistically ambitious so that one can hardly be confident about their feasibility in the foreseeable future.

Therefore, reform towards a more centralised approach to water rights administration seems less politically infeasible in China. In other words, as agencies for the MWR, river/lake basin authorities are less likely to take over authorities in water rights administration that are currently vested in provincial WRBs or WABs, at least in the foreseeable future. A more suitable solution might be keeping the parallel approach to water rights administration (based on both river basins and administrative units), and continuing to strengthen basin authorities by clarifying separation of functions between river basin authorities and provincial governments.

7.4.1.2 Is Centralised Water Rights Administration at the Provincial Level Possible?

Similar to the central-provincial relationship, division of authority among provincial, prefecture, and county governments is far from formally specified in China.[185] Instead, their relationships are negotiated on a case-by-case basis and subject to potential readjustments to the changed conditions and needs. As a consequence, one can find decentralisation and recentralisation simultaneously taking place in China to divide authority among local governments. In addition, fiscal and institutional arrangements below the provincial level are varied in different areas of China as a result of its subnational structure allowing provincial governments to determine the specific arrangements in their local contexts.[186] One example of decentralisation in recent times is that more powers for economic and social management have been transferred from provincial governments and prefecture governments to county governments. This transfer of authority among the three local level governments is also associated with a fiscal and institutional reform below the provinces, also known as the "Province-Managing-County" fiscal reform.[187] Within this context, rural-based counties, which used to be subordinate to urban-based prefectures for their fiscal budget and personnel appointment, have become a government level equivalent to prefectures. In essence, the reform involved an institutional change in

---

[184] Bin Liu, "Institutional Design Considerations for Water Rights Development in China," in *Water Rights Reform: Lessons for Institutional Design*, ed. Bryan Randolph Bruns et al. (Washington, D.C.: International Food Policy Research Institute, 2005), 275–276.

[185] Starr, *Understanding China*, 71. Also see World Bank, *East Asia Decentralizes*, 11, 31.

[186] World Bank, *East Asia Decentralizes*, 11.

[187] Yongzheng Liu and James Alm, "'Province-Managing-County' Fiscal Reform, Land Expansion, and Urban Growth in China," *Journal of Housing Economics* 33 (2016): 82.

vertical subnational government administration, through which the previous "province-prefecture-county" framework has been transformed into a "province-prefecture and county" framework.[188] As a result of reduction in governmental layers, counties have obtained more fiscal resources and decision-making autonomy to help speed up economic growth in rural areas and reduce income gaps between rural and urban areas. By the end of 2009, 22 provincial regions had commenced the reform on a pilot basis.[189] Pilot reforms indicated that decentralisation effectively provided development impetus for county economies, and successfully increased counties' fiscal capacities.[190] The reform was then further promoted by the central government for wider implementation.[191] By the end of 2011, the reform had been widely carried out across China, except in a few provincial regions (mostly remote and minority autonomous regions).[192]

Interestingly, China's decentralisation is also coupled with some examples of recentralisation that tend to occur in national resources administration. One example is that the authority in land use administration has been recentralised to the provincial level. This is mainly a policy response to the growing problems that the central government faced in monitoring the activities of local governments. As noted earlier, local governments, particularly those at lower levels, often prioritise local economy over environmental protection, and therefore, they "can thwart central directives by inaction".[193] In many cases, local governments violate regulations on natural resources management. As land lease presented as a major revenue earner for local governments, urban land expansion was induced as a result of the fiscal reform.[194] In response to these issues, the State Council initiated a land use administration reform below the provincial level, under which the overall land use plans at sub-provincial levels have to be assessed and approved by the provincial governments.[195]

Based on the above analyses regarding China's decentralisation and recentralisation practices, it can be argued that there are some indications of recentralisation in natural resources administration, at least within provincial units, though other administrative powers with respect to economic activities show an adverse trend.

---

[188] China has a four-level framework of subnational governments, which includes provinces, prefectures, counties and townships. Township governments, however, tend to have few powers for the administrative permits and approvals on various social and economic activities. Instead, these administrative powers are often divided among provincial, prefecture and county governments.

[189] Liu and Alm, "Province-Managing-County' Fiscal Reform," 85.

[190] Ibid.

[191] Ibid.

[192] Ibid, 88.

[193] Starr, *Understanding China*, 72.

[194] Liu and Alm, "Province-Managing-County' Fiscal Reform," 83.

[195] 2004 China State Council's Notice Regarding the Reform of the Land Use Administration System below the Provincial Level, http://www.cas.ac.cn/html/Dir/2004/12/01/7810.htm (accessed September 29, 2008). The administration of overall land use plans of townships may be delegated to prefectural governments, but several major indicators in the plans are still subject to the assessment and approval of the provincial government.

Without formal division of authority, the relationships among central, provincial, and sub-provincial governments are constantly evolving as circumstances change over time. On one hand, the economic reforms have largely expanded subnational activities and thus led to further decentralisation. On the other hand, however, while local governments become independent interest groups who pursue their own economic interests, decentralisation is likely to jeopardise public values of the society such as equality and environmental protection due to deficiencies in monitoring local governments' activities. Recentralisation then becomes a logical response.

In the case of land resources, recentralisation has been employed to address the problem of illegal land use by local governments. Back to the case of water resources that face a similar problem to land resources, one could reasonably argue the possibility for water administration to be recentralised within provincial units. Moreover, recentralisation can serve not only as a prescription to the monitoring problem, but also as a facilitating institutional arrangement to remove administrative barriers for water rights trading. If water rights administration were recentralised at the provincial level, the intra-provincial water rights trading market may be easier to grow.

To conclude, it is recommended that China's water rights administration should be recentralised at both the provincial level and at the river basin level. Because it is politically difficult to have full centralisation at the river basin level, the present institutional arrangement which allows the water rights administration power to be shared between river/lake basin authorities and provincial WRBs or WABs may be maintained and further clarified. Recentralisation is more feasible within provincial units. In order to facilitate water rights trading, provincial WRBs or WABs with centralised authority in water rights administration may overperform a decentralised approach.

7.4.2 *Water Rights Titling/Registration System*

How a water rights market operates relies on the extent to which the water rights titling/registration system provides security of title and convenience in transactions. Jurisdictions often maintain a database of existing water rights in their permitting authorities, and some also have a separate organisation to take such function. The difference between "the registers of rights" and "the registers of deeds" implies distinctions in the level of title security and transaction costs for water transfers, and thus affects the operation of the water market.

In the absence of proper institutional arrangements enabling water trading, registration issues are less addressed in the current water rights administration system in China. Water rights transfers remain predominantly driven by government, and a water rights market in which private water rights holders actively participate and seek voluntary trading parties is yet to emerge. Most water transfer transactions take place between governments (represented by irrigation district authorities) and new water users, in which cases the verification of water rights does not present as an essential part of the trade. Accordingly, policy makers seemed to have paid less attention to the titling and registration issue. However, in order for China to utilise water rights trading for better water allocation, its current water rights registration system needs to be refined so that the title of water rights can be secure enough and transaction costs low enough to encourage and facilitate trades. Currently, local governments have been amending their water regulations according to the new provisions regarding water abstraction permits. Water abstractors are required to have their old permits updated or have their historical abstraction permitted and registered in their local water authorities. Therefore, China's water rights registers are yet to be completed. It needs to go through this process before a water rights registration system can be established.

Consistent with the proposed centralisation of water rights administration at the provincial level, registers of water rights should also be centralised from sub-provincial water administrators (county and prefecture WRBs or WABs) to the provincial WRBs or WABs. Through this recentralisation of water rights administration and registration, a province-wide market could potentially occur which enables water abstraction right holders to search sources of water rights from different counties or prefectures. Local WRBs or WABs at the county and prefecture levels should cooperate with the provincial WRBs or WABs to pursue recentralisation of water rights registration as a step-by-step approach. As water rights administration gets recentralised at the provincial level, county and prefecture WRBs or WABs should no longer take charge of examining and approving new applications for water abstraction permits. Rather, they should keep involved in conversion of existing old water abstraction permits or historical abstractions without permits into the updated water abstraction permits in accordance with new regulatory requirements.

Continuing to be responsible for filing registers of existing water abstractions, county and prefecture WRBs or WABs provide information about the existing registers to provincial agencies and assist them to build up a province-wide registry. Similar cooperation should also be carried out between provincial WRBs or WABs and river/lake basin authorities to build up a basin-wide registry. The provincial registers that maintain their own water rights registers in accordance with their jurisdictions should also report to river/lake basin authorities. Although currently there is no market transfer taking place between provinces except the SNWDP, registers should be collected together at the river basin level to prepare a pool of water rights for future expansion of the water market.

The registration system of water rights should also take the indefeasibility of title into consideration. According to international experiences, the registration system of water rights in a given jurisdiction usually follows a pattern adopted from its land title registration system. Similarly, it would be useful for China to refer to the registration system for real estate titles to establish its water rights registration system. The new Property Law of China has refined its registration system for real estate titles. Provisions in the Property Law regarding the titling and registration of real estate titles include the following:

- The creation, alteration, alienation or termination of the real right of a realty shall come into effect as it is registered in accordance with law. Unless it is otherwise prescribed by laws, it shall have no effect if it is not registered in accordance with law. As regards the ownership of the natural resources owned by the state in accordance with law, the registration is not required. (Article 9)[196]
- The registration of a realty shall be dealt with by the registry where the realty has been located (Article 10). Administered by the registry, the realty register shall be the basis for determining the ownership and contents of a realty (Article 16). Any right holder or interested party may file an application for consulting or copying the registration information, and the registry shall provide it (Article 18).[197]
- The realty ownership certificate shall be the evidence for the holder's ownership of a realty (Article 17).[198]
- The registry shall take the compensation liability in the case that damages are caused by virtue of any error in registration. After paying the compensation, the registry may recover the amount from the person who has caused the registration error (Article 21).[199]

Based on the above listed provisions, the titling/registration system for real property in China falls in the category of "registers of rights", which gives the right holders indefeasible rights. With government guarantee available for right holders, potential investors may be more confident in the certainty of title and security of transaction. Rather than having to search historical transactions related to a certain title, they can simply rely on the information recorded in the registry. This registration system applying to real properties in China should also be applicable to water usufructuary rights that are recognised as a type of real right in the new Property Law.

In addition to the registration of water usufructuary rights, it is worth noting that an effective registration system is essential for the SNWDP. Although provincial governments, their water suppliers and end water users are not conferred a usufructuary right, their water entitlements, whether shared-based or contractual, should be specified and registered in a relevant route corporation to underpin the future water trading market in the project.

[196] 2007 Property Law of the People's Republic of China, http://www.npc.gov.cn/englishnpc/Law/2009-02/20/content_1471118.htm (accessed May 21, 2017), art. 9.

[197] Ibid, art.10, art. 16, and art. 18.

[198] Ibid, art. 17.

[199] Ibid, art. 21.

7.4.3 *Regulation of Water Rights Trading Markets and Provision of Trading Platforms*

Policy makers in China have highlighted the "quasi market" characteristic of its water rights trading based on their concern about government control over water resources. Although it is consistent with China's traditional approach to water governance, the concept of "quasi market" may be misleading with its vogue meaning. Does it refer to a market allocation institution with heavy government intervention, or a market-like government allocation institution? According to the current practices in the Yellow River Basin, it is more like the latter, which seems to be inconsistent with the purpose of China's ongoing water policy reform. As the reform deepens, the market forms of water rights trading need to be further explored.

At present, there is a lack of "water rights trading market" in a strict sense in China. Trading platforms are yet to be provided for water rights trading. Water notes transfers between farmers in Zhangye can be considered as water trading market, but what has been transferred is water supply contracts rather than water usufructuary rights (water abstraction permits). Water rights transfers from agriculture to industry in the Yellow River Basin do involve transfer of water abstraction permits, but it has been driven by government on a planning basis. Some have proposed the establishment of a water bank for the SNWDP, but again, it is still a water market mainly for water supply contracts rather than usufructuary rights. As part of the nation-wide water rights piloting programme commenced in 2014, four provincial regions (Inner Mongolia, Henan, Gansu, and Guangdong) were selected to explore water rights transfer models, in which the establishment of a water rights trading platform was identified as a focus. The 2016 Provisional Measures on Administration of Water Rights Trading requires that water rights trading should usually take place on water rights trading platforms, and may also take place directly between transferor and transferee.[200] For regional water rights trading and water abstraction rights trading with a major water amount, trading platforms should be used to provide trading parties with related trading services.[201]

A few months later after the release of the 2016 Provisional Measures on Administration of Water Rights Trading, China launched its national water exchange in Beijing. Jointly founded by the MWR and the Beijing Municipal Government, China Water Exchange will facilitate water rights trading, and provide services including trading consultation, technical evaluation, market information, intermediary services, and public services.[202] According to the trading information published on the China Water Exchange website, there have been a total of 15 trade completions between June 2016 and March 2017, among which 11 are regional water rights

---

[200] 2016 Provisional Measures on Administration of Water Rights Trading, art. 7.

[201] Ibid.

[202] Information from the official website of China Water Exchange, http://cwex.org.cn/html/cn/ (accessed May 25, 2017).

trades and water abstraction rights trades, with four irrigation rights trades.[203] China Water Exchange developed a range of trading rules, including the 2016 China Water Exchange Water Rights Trading Rules (Trial).[204] While trading parties can negotiate about the trade before they finalise the trade on the trading platform, China Water Exchange also matches buyers and sellers through a bulletin board. The effectiveness of the newly established China Water Exchange in facilitating water rights trading is yet to be evaluated.

In addition to the Beijing-based national water exchange, local based water markets serving specific regions should also be developed for facilitate trading in water abstraction rights. Corresponding to the proposed two level framework of water rights administration, China could establish two levels of water markets for trading in water abstraction rights, consisting of river basin water markets and provincial water markets. River basins here refer to trans-provincial ones with national importance in China such as the Yellow River and the Yangtze River, and river basin markets have already reached their broad scope to internalise water users likely to be interested in water rights trading. Restricted by the physical connections of the hydrological systems, water rights trading tends to take place within river basins or smaller water systems. With the SNWDP in place, China's four major river basins will be connected into an integrated water system so that the interbasin transfers become physically possible. Subject to the trading rules established by the route corporations of the SNWDP, interbasin transfers of water supply contracts should be allowed between the project contractors. As for water abstraction rights other than the project water entitlements, interbasin transfers (trans-provincial ones) may be restricted by the provincial governments where the permit has been issued in order to maintain their control over their total water quantity quotas.

7.4.4 Protection of Third-Party Interests

Because of water's unique characteristics as both an economic and social good, its market allocation may need higher a level of government intervention to address externality issues and potential market failures, which are associated with the nature of the water cycle and interdependence among different users. As values of water keep changing, the water rights trading market is developing towards a new direction, which is not only working for resource allocation efficiency but also for resource sustainability. Third-party interests have been interpreted in a broader sense to incorporate increasing awareness of social, cultural, and environmental values of water resources that all call for accommodation in the market mechanism. Reflecting changing values of water, the developments of overseas water rights

[203] Information from the official website of China Water Exchange, http://cwex.org.cn/lising/ (accessed May 25, 2017).

[204] 2016 China Water Exchange Water Rights Trading Rules (Trial), http://cwex.org.cn/2016/jygz_0621/45.html (accessed May 25, 2017).

trading markets have clearly shown diligent efforts in protection of third-party interests. These efforts have been made from two different standing points. One is minimising the potential negative effects upon third parties by setting up protective barriers such as trading restrictions and approval procedures. The other is directly utilising the market mechanism to ensure public values, through which government or private organisations interested in the environmental and ecological protection of water systems can participate in the market as transaction parties and purchase water rights for instream uses.

Despite substantial developments in regulation of water trading markets, third-party interests remain less addressed in China. In many cases of current water transfer practices, governments have participated in the transactions as sellers of water rights, and farmers whose water use has been reduced as a result were neither involved in decision making nor compensated for their reduced water availability. This fundamentally resulted from the absence of legally recognised farmer level water rights under the traditional government allocation institution. Therefore, to effectively protect third-party interests in water market transfers, the first step for China is to legitimately recognise and clearly specify water rights at lower levels including those of WUAs and farmers. With specified and enforceable water rights in place, the water rights administration system then should incorporate water transfer restrictions or approval processes so that mechanisms are available for third parties to defend their legal interests.

Similarly, the environment is yet to be valued as a third party for whom protection mechanisms are barely available in China. Although sustainable water use has been explicitly acknowledged in law and policy, economic growth is still a higher priority in reality, especially for local governments and communities. Indeed, the current water reform tends to ensure the continuing rapid economic growth without worsening the environmental deterioration, which implies its preference of economy over the environment. Within this context, one can hardly expect that environmental values are carefully taken into account in water rights trading. Accordingly, it can be argued that a water market for the environment is not likely to take place in China. Nevertheless, one should be aware that the market mechanism of water rights trading does effectively help reduce and control off stream uses and indirectly protect the value of water for the environment. From this perspective, the extent to which the mechanism works for the environment is more dependent on the determination of the water volume available for abstraction, which is the task of water planning. The dilemma of development and environmental protection remains the most critical issue that China has to deal with along its water reform.

7.5 Concluding Remarks

The water rights administration system is of great significance in facilitating the water rights trading market. It administers the issuance of new rights, approves changes and transfers of existing rights, and provides security of title for right holders through its registration system. How these aspects are arranged in law and institution have major implications for the market performance. As China's current legal system has just begun its transformation from the traditional administrative control over water use, it has not yet developed a sophisticated administration system for water rights trading. A reasonable number of provisions has been developed, available to govern conservation-based water rights transfers. The development of trading platforms is still in its infancy, as the newly launched China Water Exchange is expected to play a leading role. Rather than operating as a market where buyers and sellers voluntarily search out transaction parties and reach agreements, water rights transfers in China remain heavily driven by government, and substantially controlled based on a long-term planning horizon. Due to these constraints, the water rights administration system lacks sufficient institutional arrangements to facilitate wider implementation of the market mechanism.

This chapter has suggested several reform recommendations with respect to the administration system of water rights trading based on the examination of the current system of China and reflections on overseas experiences. They can be summarised as follows:

- Due to the division of powers in issuing water abstraction rights and decentralised water rights administration at multiple subnational levels, the present institutional framework tends to create administrative barriers to water rights trading and limit it within the small geographical scope of county and prefecture units. With political feasibility in mind, it is proposed that water rights administration should be recentralised at provincial and river basin levels to allow water rights trading to take place within broader scopes.
- Water rights titling/registration systems should be established at two levels: provinces and river basins. This is consistent with the recentralisation of the water rights administration system. To provide security of title sufficient to allow greater public and investor confidence and lower transaction costs in the verification of water rights, a system of "registry of rights" applied to real estate titles should also be applied to water usufructuary rights.
- Similarly, water marketplaces could also be formed at two levels (provinces and river basins) corresponding to the water rights administration system. Basin-wide water markets may be difficult to form at the current stage of incomplete water quantity allocation between provincial regions. Alternatively, pilot projects of provincial water markets may be initiated to determine the suitability of specific market forms within different local contexts of provinces. Procedures of temporary water rights transfers should be streamlined in some market form to ensure speed and efficiency.
- Although conservation-based transfers help reduce the risks of negative third-party effects, direct government involvement in transactions as water rights transferors potentially involve conflicts of interest between government agencies. Third parties concerned about water rights transactions, especially farmers whose water supply may be affected by intersectoral water transfers between

irrigation districts and industries should be provided participation mechanisms to defend themselves against potential negative effects.
- Wider third-party interests such as those in the environment can hardly be accommodated in China's water rights trading market due to the fact that the government's as well as the public's preference is still given to economic rather than environmental protection. Notwithstanding that instream water transfers are unlikely to take place in China's water rights trading market, the market can effectively help government to control the volume of water use and improve water conservation provided that water plans, and particularly the total water volume available for consumptive uses, can be determined in a sustainable manner.

By proposing recommendations to reform the current system for water rights administration, this chapter has argued that government intervention in the water rights trading market should be modified in order to allow the mechanism to function well. As water abstraction right holders, irrigation district authorities should achieve further separation from water administration authorities to avoid governments being both "gamekeeper" and "poacher" and allow governments to focus on regulation and compliance. Rather than directly participating in the market as water rights holders, governments should be more responsible for providing sufficient institutional infrastructures indispensable for the operation of the water trading market. To be specific, this responsibility involves maintaining registries of water rights, articulating trading rules for water rights transfers, and providing trading platforms for buyers and sellers.

References

Abeln, R. (2005). Instream flows, recreation as beneficial use, and the public interest in Colorado Water Law. *University of Denver Water Law Review, 8*, 517–545.

ACIL Tasman, & Freehills. (2004). *An effective system of defining water property titles*. Research Report to the Australian Government Department of Agriculture, Fisheries and Forestry, and Land & Water Australia. Canberra: Land & Water Australia.

Australia National Water Commission (NWC). (2011). *Water markets in Australia: A short history*. Canberra: National Water Commission.

Australia National Water Commission (NWC). (2013). *Australia water markets report 2012–13*. Canberra: National Water Commission.

Australian Bureau of Agricultural and Resource Economics and Sciences (ABARES). (2016). *Australian water markets report 2014–15*. Canberra: Department of Agriculture and Water Resources.

Australian Competition and Consumer Commission (ACCC). (2010). *Water market intermediaries – Industry developments and practices*. Canberra: Australian Competition and Consumer Commission.

Australian Productivity Commission. (2003). *Water rights arrangements in Australia and Overseas*. Melbourne: Productivity Commission.

Bauer, C. J. (1998). *Against the current: Privatization, water markets, and the state in Chile*. Boston/London: Kluwer Academic.

Bonham, C. H. (2006). Perspectives from the field: A review of western instream flow issues and recommendations for a new water future. *Environmental Law, 36*, 1205–1235.
California State Water Resources Control Board. (2017). *Water rights applications: Permitting and licensing program*. http://www.swrcb.ca.gov/waterrights/water_issues/programs/applications/. Accessed 2 Mar 2017.
China Ministry of Water Resources, Water Management Department. (2005). *The report on pilot experiences of establishing the water rights system part 1: Materials of water rights transfer in the inner Mongolia autonomous region*. Beijing: China Water Publishing Co..
Colorado Division of Water Resources. (2012). *Guide to Colorado well permits, water rights, and water administration*. http://water.state.co.us/DWRIPub/Documents/wellpermitguide.pdf. Accessed 22 May 2017.
Dyckman, C. S. (2005). A dynastic disruption: The use efficiency and conservation legacy of the Governor's Commission to review California Water Rights Law Recommendations. *McGeorge Law Review, 36*, 175–208.
Getches, D. H. (1997). *Water Law in a Nutshell* (3rd ed.). St. Paul: West Publishing Co..
Gray, B. E. (2008). The market and the community: Lessons from California's Drought Water Bank. *Hasting West-Northwest Journal of environmental Law & Policy, 14*, 41–102.
Grow, D. A. (2010). Policy punctuations in Colorado Water Law: The breakdown of a monopoly. *Review of Policy Research, 27*(2), 147–166.
Haddad, B. M. (2000). *Rivers of gold, designing markets to allocate water in California*. Washington, DC: Island Press.
Hanak, E., & Dyckman, C. (2003). Counties wresting control: Local responses to California's statewide water market. *University of Denver Water Law Review, 6*, 490–518.
Hennessy, M. (2004). Colorado river water rights: Property rights in transition. *University of Chicago Law Review, 71*, 1661–1687.
Hodgson, S. (2006). *Modern water rights: Theory and practices*. Rome: Food and Agriculture Organization of the United Nations.
Howe, C. W. (1997). Increasing efficiency in water markets: Examples from the Western United States. In T. L. Anderson & J. Peter (Ed.), *Water marketing – The next generation*. Hill/Lanham/Boulder/New York/London: Rowman & Littlefield Publishers, Inc..
Jones, P. A., & Cech, T. (2009). *Colorado water law for non-lawyers*. Boulder: University Press of Colorado.
Kaiser, R. A. (1996). Texas water marketing in the next millennium: A conceptual and legal analysis. *Texas Tech Law Review, 27*, 181–261.
King, M. A. (2004). Getting our feet wet: An introduction to water trusts. *The Harvard Environmental Law Review, 28*, 495–534.
Koehler, C. L. (1995). Water rights and the public trust doctrine: Resolution of the Mono Lake controversy. *Ecology Law Quarterly, 22*, 541–589.
Liu, B. (2005). Institutional design considerations for water rights development in China. In B. R. Bruns, C. Ringler, & R. Meinzen-Dick (Ed.), *Water rights reform: Lessons for institutional design* (pp. 275–276). Washington, DC: International Food Policy Research Institute.
Liu, Y., & Alm, J. (2016). 'Province-Managing-County' fiscal reform, land expansion, and urban growth in China. *Journal of Housing Economics, 33*, 82–100.
Neuman, J. C. (2004). The good, the bad, and the ugly: The first ten years of the Oregon Water Trust. *Nebraska Law Review, 83*, 432–484.
New South Wales Department of Primary Industries Water. (2017). *Water licensing*. http://www.water.nsw.gov.au/water-licensing. Accessed 2 Mar 2017.
Rose, C. M. (1998). A review of Professor Joseph Sax's defense of the environment through academic scholarship: Joseph Sax and the Idea of the Public Trust. *Ecology Law Quarterly, 25*, 351–362.
Sax, J. (1970). The public trust doctrine in natural resource law: Effective judicial intervention. *Michigan Law Review, 68*, 475–566.

Scoot, A., & Coustalin, G. (1995). The evolution of water rights. *Natural Resources Journal, 35*(4), 821–979.
Smith, G. P., & Sweeney, M. W. (2006). The public trust doctrine and natural law: Emanations within a Penumbra. *Boston College Environmental Affairs Law Review, 33*(2), 307–343.
Starr, J. B. (1998). *Understanding China*. London: Prfile.
Thompson, B. H. (1993). Institutional perspectives on water policy and markets. *California Law Review, 81*, 673–764.
Weber, G. S. (1995). Articulating the public trust: Text, near-text and context. *Arizona State Law Journal, 27*, 1155–1248.
Woolston, M. (2005). Registration of water titles: Key issues in developing systems to underpin market development. In J. Bennett (Ed.), *The evolution of markets for water: Theory and practice in Australia*. Cheltenham/Northampton: Edward Elgar.
World Bank. (2005). *East Asia decentralizes: Making local government work*. World Bank: Washington, DC.

Legal and Policy Sources

1989 Administrative Procedure Law of People's Republic of China. http://www.npc.gov.cn/englishnpc/Law/2007-12/12/content_1383912.htm. Accessed 22 May 2017.
1993 Implementing Measures on the Water Abstraction Permits System (China State Council). http://www.chinawater.net.cn/law/W03.htm. Accessed 28 Sept 2008.
1999 Law of People's Republic of China on Administrative Reconsideration. http://www.npc.gov.cn/englishnpc/Law/2007-12/11/content_1383562.htm. Accessed 22 May 2017.
1999 MWR Interim Rules on Administrative Reconsideration (China Ministry of Water Resources). http://www.sxwater.gov.cn/oldweb/zcfg/SLXZCFFLFG/SLXZCFFLFG1019.htm. Accessed 28 Sept 2008.
2000 Water Management Act of New South Wales. http://www.austlii.edu.au/au/legis/nsw/consol_act/wma2000166/. Accessed 22 May 2017.
2003 Administrative Permission Law of the People's Republic of China. http://www.npc.gov.cn/englishnpc/Law/2007-12/05/content_1381966.htm. Accessed 22 May 2017.
2004 China State Council's Notice Regarding the Reform of the Land Use Administration System below the Provincial Level. http://www.cas.ac.cn/html/Dir/2004/12/01/7810.htm. Accessed 29 Sept 2008.
2004 Trial Implementing Measure on Administration of Water Rights Transfer in the Yellow River Basin (Yellow River Conservancy Commission). http://www.yrcc.gov.cn/ziliao/zcfg/guizhang/200612/t20061222_9390.htm. Accessed 26 Sept 2008.
2006 Regulation on Administration of Water Abstraction Permits and Water Resources Fee Collection (China State Council). http://www.gov.cn/zwgk/2006-03/06/content_220023.htm. Accessed 21 May 2017.
2007 Property Law of the People's Republic of China. http://www.npc.gov.cn/englishnpc/Law/2009-02/20/content_1471118.htm. Accessed 21 May 2017.
2008 Measures on Administration of Water Abstraction Permits (China Ministry of Water Resources). http://www.gov.cn/flfg/2008-04/17/content_947055.htm. Accessed 21 May 2017.
2016 China Water Exchange Water Rights Trading Rules (Trial). http://cwex.org.cn/2016/jygz_0621/45.html. Accessed 25 May 2017.
2016 Provisional Measures on Administration of Water Rights Trading (China Ministry of Water Resources). http://www.gov.cn/zhengce/2016-05/22/content_5075679.htm. Accessed 27 May 2017.

Websites

The official website of China Water Exchange. http://cwex.org.cn/html/en/. Accessed 25 May 2017.

The official website of Ministry of Environmental Protection of China. http://www.zhb.gov.cn/hjjc/hjzf/djt/200612/t20061229_99016.htm. Accessed 28 Aug 2008.

The official website of the Department of District Division and Place Naming, China Ministry of Civil Affairs. http://www.xzqh.org.cn/ONEWS_zq.asp?id=1150. Accessed 28 Aug 2008.

The official website of the Inner Mongolia Department of Water Resources. http://www.nmgslw.gov.cn/info/infoView.jsp?idcontent=9. Accessed 14 July 2008.

Chapter 8
Conclusion

Abstract This chapter is the concluding chapter of this book. It first provides an overview of the theoretical enquiry made in the book by presenting a water rights governance spectrum. It also draws together the proposals that are made throughout the book to feed pragmatic strategies for China's water law and policy reform to move towards tradable water rights. It puts China's water reform in the context of its wider societal transition, as well as the wider context of international water governance discussions and policy challenges in responding to future changes in water supply and demand. It also points out directions for future research.

Keywords Water rights governance spectrum • Government/market balance • Tradable water rights • Reform recommendations • Third party protection

8.1 Overview

Water links multiple users, including humans and the environment, with each other in its integral hydrological cycle. The mobile and fugitive nature of water complicates the interdependency of co-existing water users. In the context of various existing stressors (economic development, growing population, lifestyle changes, and changing values of ecosystem services) coupled with climate change, how water governance responds to those challenges is at the centre of dealing with water scarcity, variability and uncertainty. The unique characteristics of water resources and the complexity of water use have significant implications for shaping water governance. As water scarcity worsens in many places around the world, water governance is increasingly recognised as a central issue to tackle water resource challenges. Water allocation mechanisms deal with multiple values of water and set the rules for water use, lying at the heart of water governance. Against this backdrop, the world has seen many countries undertaking major reforms in water governance for more sustainable water management. Jurisdictions across both common law and civil law systems have been implementing major changes to the way water is allocated and reallocated, including the adoption of tradable water rights and development of water trading markets.

M. Jiang, *Towards Tradable Water Rights*, Global Issues in Water Policy 18,
DOI 10.1007/978-3-319-67087-4_8

In response to its water crisis, China is one of these countries that have attempted to introduce the water rights trading system into its water law and policy. Leading the world's economic growth, China's development has seriously overstretched its natural resources. It is not surprising to see that China is facing one of the world's most severe water crises.[1] China's water crisis involves water shortage, pollution, and ecosystem deterioration, each presenting distinctive challenges for policy reform. China's water governance is therefore increasingly challenged to balance multiple and diverse values of water use. Within this context, an ambitious water reform, "Building a water saving society", has been initiated and undertaken by the Chinese Central Government since 2002.[2] The water reform aims to integrate the traditionally fragmented water management systems, establish a national water rights system, and implement market-based water allocation mechanisms for more sustainable water use.[3]

Whereas a preliminary framework of national water rights system has now been established after over 10 years into the reform, water use security as well as water quality control remain immense challenges for China. In early 2012, the Chinese Central Government pushed the water reform forward towards more strict water governance, with the intention to refine its institutional arrangements for total quantity control of water consumption and quota management.[4] The reform implies that where the total quantity of water consumption and abstraction has reached or exceeded the defined total limit, no further water abstraction permits should be issued or approved. Water rights trading then becomes an alternative vehicle to obtain water use rights. China's water reform therefore calls for further development of market-based water allocation mechanisms, refinement of the water rights system, establishment of water markets, and encouragement and facilitation of water trading.[5]

Water rights trading has been both theoretically and empirically proven as an effective and efficient resource allocation instrument within certain contexts. In order for it to function well, some indispensable elements have to be in place. Observations on international experiences in water rights trading policy and practice, with a particular focus on those in Australia, suggest that enabling factors for a well-functioning water market include: clearly defined tradable water rights, sophisticated titling/registration system, established market regulation, and access to trading platforms and intermediaries. In addition, broad institutional and governance

---

[1] 2012 China State Council Opinions on the Implementation of the Strictest Water Resources Management System, State Council Document No.3, http://www.gov.cn/zwgk/2012-02/16/content_2067664.htm (accessed October 24, 2016).

[2] Ibid.

[3] Shucheng Wang, *Resource-oriented Water Management: Towards Harmonious Coexistence between Man and Nature,* 2nd ed. (Singapore: World Scientific Publishing; Beijing: WaterPower Press, 2006).

[4] 2012 China State Council Opinions on the Implementation of the Strictest Water Resources Management System.

[5] Ibid.

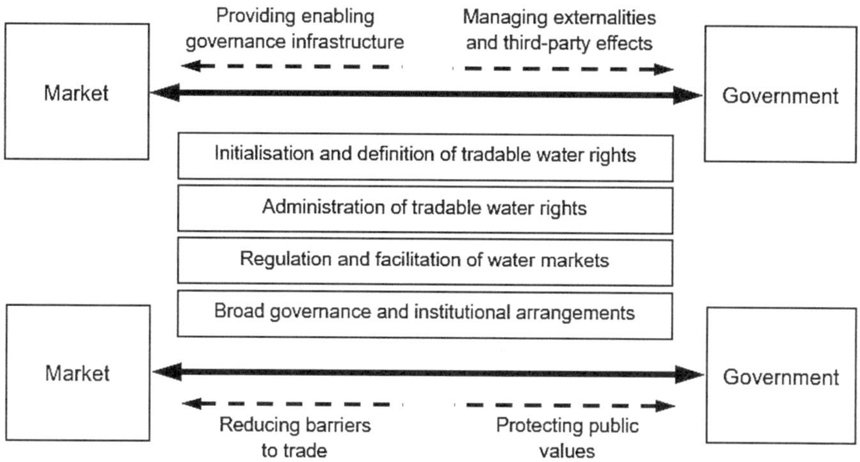

Fig 8.1 Water rights governance spectrum

arrangements are also important to underpin functioning of water markets. These factors contribute to either providing essential governance infrastructure for the development of water markets, or reducing barriers to water trading and encouraging market participation. Evolution in water rights and markets involves adjustments and movements between different institutional arrangements for water allocation, highlighting the wrestle between government and market. Put in another way, there is a need for balanced integration of government intervention and market mechanisms. Indeed, how to effectively address externalities and market failures, and facilitate the development of water markets at the same time, remains a central issue of the water rights system. The above figure (Fig. 8.1) illustrates the water rights governance spectrum in which the choice of institutional arrangements depends on different policy objectives, thereby involving different levels of government intervention in water markets.

China's water law and policy reform towards tradable water rights starts at an extreme administrative approach to water allocation. This tradition in high level of government control as its starting point has significant implications for what directions and priorities China will take towards its water reform. How can water markets work well with adequate but not overreaching government intervention within China's context? How can China achieve a balance between market freedom and government intervention? These questions are central to the reform. It is hoped that this book has contributed to exploring answers to these important questions. In particular, it has examined the deficiencies of the current systems for water rights arrangements and trading in China, and proposed potential directions for future development in water rights law and policy.

8.2 Summarising the Key Messages

The book has set its focus on two major aspects of government involvement in the market mechanism of water rights trading, which consist of two essential parts of China's national water rights system currently under construction, namely the initialisation and definition of tradable water rights and administration of water rights and trading. China's current practices in water rights trading and other water allocation instruments have been assessed by examining several representative examples and pilot cases under its ongoing water policy reform. It has been observed that water rights trading has indicated its effectiveness by improving efficiency, equity, and sustainability of water resources allocation, while other instruments including the technological instrument of developing new water supplies, the traditional administrative allocation, and water pricing have tended to show limitations. However, the empirical analysis has shown that the functioning and the wider implementation of water rights trading have been hindered by the current legal and institutional system which originally works for government control over water, and therefore lacks sufficient infrastructure to underpin and facilitate the market mechanism. Several crucial impediments include ill definition of water rights, poor performance of water rights administration, and failure in monitoring and enforcing water rights. These constraints all relate to the issue of how to balance and mix government and market in China's water allocation institution.

The book has made reform proposals for China to define tradable water rights with adequate security and to refine its water rights administration to facilitate and underpin the water rights trading market by analysing and assessing China's current legal systems and referring to international experiences from overseas jurisdictions. Among other jurisdictions, Australia has been a particular focus where the development of water markets is recognised internationally as a water reform success story.[6] This comparative approach allows feasibility of the recommendations to be carefully examined within China's present legal and institutional contexts.

China has been working hard to initialise water rights and make them tradable for more efficient and sustainable water allocation. Along with changes in the philosophy of water use towards sustainability, private property rights to water have begun to acquire both legal and public recognition. A process from water planning through water quantity allocation to water abstraction permits divides usufruct rights from state ownership over water resources. At the macro level, water planning and water quantity allocation are underway to limit total water use and to clarify regional water quotas. At the micro level, water abstraction permits are converted from administrative licences to water titles. Nevertheless, the definition of tradable water rights in China is still in its infancy. The empirical analysis has indicated that ill-defined water rights provide little security to water users, which discourages

[6] DebraTownsend and Odette Adams, "Water Rights and Trading in Australia," *Australian Environmental Law Digest*, September (2016): 3-4. K. William Easter and Qiuqiong Huang, eds. *Water Markets for the 21st Century: What Have We Learned?* (Dordrecht: Springer, 2014), 46.

trading. To summarise, the book has clarified several essential aspects of the present legal framework for water abstraction rights, and made reform recommendations to improve rights security. With better understanding of the approaches to water rights arrangements and the legal nature of regional water rights of local governments, it has concluded that lower level water rights, including irrigation water rights of WUAs and individual farmers (in the form of the contractual rights arrangement), should be explicitly recognised by statutes and clearly specified. To exert its function in water allocation efficiency and flexibility, the divisibility and transferability of the water rights should be further augmented to enable temporary water rights transfers. Consistent with these recommendations, the book has also introduced a proposal to reform the institutional framework for China's potentially largest water rights trading market, the SNWDP. Unlike the MWR's proposal, which failed to fully recognise the usufruct rights, the new proposal discussed in this book has explicitly recognised water usufruct rights of the corresponding corporations involved in the Middle Route and the Eastern Route projects, and share-based water entitlements of provincial governments. These water rights holders as well as water supply contractors of the SNWDP should be given clearly defined and tradable water rights in order for the market to grow.

Besides exploring the initialisation and definition of tradable water rights, the book has further addressed water rights administration, which underpins the water rights trading market. It has been observed that the current water rights administration system in China lacks sufficient institutional arrangements to facilitate wider implementation of water rights trading. Water rights are administered in a decentralised manner as multiple levels of government share the power of water rights administration. A water rights registration system is also yet to be established in China. In July 2014, the MWR commenced its nationwide water rights piloting programme, which includes water rights verification and registration as a major focus. Based on the experiences learnt through the programme, the MWR will develop regulations and guidelines on water rights verification and registration. The 3-year piloting programme is expected to complete by the end of 2017. In this context, effective monitoring and enforcement arrangements are not in place to ensure compliance of water laws and regulations, for which the value of water rights is compromised. As a result, water users may ignore the limits in their water abstraction permits and choose to use as much water as they want if they take no liabilities or penalties for non-compliance.

At present, there is a lack of a "water rights trading market" in a strict sense in China. Access to trading platforms remains limited with the newly launched national water exchange. Many water rights trades are government driven, and provisions on market regulation are mainly based on conservation-based water transfers. To reflect on the past 16 years of water rights trading practices, the MWR released the Provisional Measures on Administration of Water Rights Trading in 2016. This recent policy development has established a three-level water rights trading framework (reginal water rights trading, water abstraction rights trading, and water rights trading between irrigators). General rules for these three types of water rights trading are also provided in the newly released ministerial regulation. Despite these progresses in water rights administration, third party interests remain less addressed in China. When water is transferred from agriculture to industry, farmers' interests in water use are often neglected. Similarly, the environment is yet to be valued as a third party for whom protection mechanisms are barely available.

The book has highlighted four major components regarding water rights administration: centralisation of water rights administration, water rights titling/registration system, market operations and forms, and protection of third-party interests. Recommendations for reform have also been made to refine the current water rights administration system in China. These recommendations call for modification of government intervention in water rights trading by arguing that, rather than directly participating in the market as water rights holders, governments should be more responsible for maintaining registries of water rights, articulating trading rules of water abstraction rights transfers, and providing market places for buyers and sellers.

8.3 The Way Forward

While the recent water reform in China has made impressive progress in policy development and practice, how its legal and institutional framework facilitates or hinders the application of tradable water rights remains less addressed in the existing scholarship. It is hoped that this book has provided an up-to-date reflection on the deepening water reform in China by focusing on how law and policy can facilitate the application of tradable water rights and development of water trading markets. Recognising the importance of learning from, and building on the international experiences in water rights trading policy and practice, the book has observed what has been implemented in overseas jurisdictions (particularly Australia and the US), which helped identify the major governance gaps for China to develop functioning water markets. The level of sophistication of the water markets is obviously different, which mirrors the different levels of development stages in China and other jurisdictions. However, it should be noted that the book is not intended to evaluate the market performances in China against other jurisdictions. Instead, it has focused on how China can benefit from the international experiences to address its water governance challenges in developing tradable water rights.

To facilitate active and dynamic water trading markets, China needs to further develop the relevant legal, policy, and governance infrastructure, foremost better defined and administered water rights arrangements. The management of conflicts in interest is critical as the current water trading markets are highly government driven in China. Governments play a dual role in water allocation and reallocation, examining and approving water abstraction permits on one hand, and holding and trading water abstraction permits on the other. The boundary of government involvement needs to be clearly defined to strengthen the legal status of water rights, offering a higher level of security. Local governments' roles as water administrator and water rights holder need to be further clarified to reduce risk of conflicts in interest.

China's ongoing water reform involves a transformation of its water allocation institutions. It is challenging due to its implications for changes in water management, and more importantly, changes in legal philosophy based on which water law and policy is shaped, implemented and enforced. Indeed, China's changing philosophy from "man can conquer nature" to sustainable water use is a key driving force for the evolution of its water allocation institutions. For the first time, the concept of sustainability is embedded in China's new national strategic water plan released in 2010. By setting up water use total control quotas for major river basins and provinces, the national water plan calls for further specification of water rights at lower levels.[7] Such institutional arrangement may be expected to provide a governance framework to underpin the development of water rights markets similar to Australia's "cap and trade" approach. In this sense, the market-based mechanisms have been applied not only to improve allocation efficiency but also to protect the environment and ecosystems. However, to what extent such policy acknowledgement is translated into practice remains a debatable question.

The water rights trading market cannot effectively ensure sustainable development of water resources unless its water plans, and particularly the total water volume available for consumptive uses, have been determined in a sustainable manner, and implemented strictly. As government's priority is still committed to economic development in reality, the environmental rights for water are not appropriately protected in China's current systems of water rights arrangements. And lastly, therefore, this book points out that environmental water rights and the protection of the environment as a third-party interest remains an important area for future research on tradable water rights for China. Currently, water's value for the environment is yet to be meaningfully accommodated in the water rights trading market, whereas the market mechanism to help protect the environmental needs for water has been increasingly applied in overseas jurisdictions such as Australia and western states in the US. In order for China to apply water rights trading to address water over-allocation, issues pertaining to the design of water rights for the environment need to be further addressed.

We are living in a changing world, where water governance has to deal with exacerbated uncertainty and variability in water availability. Climate change is arguably one of the most significant, controversial global issues facing both academic research and government policy. Freshwater resources are vulnerable and may be strongly affected by climate change. Water demand and use patterns may be impacted as well due to the impacts of climate change on agriculture, irrigation and drainage, hydropower, flood control, and other water-related sectors. It is therefore imperative to revisit water governance reform taking into account the needs to deal with the emerging overarching issues, such as adaptation to climate change, building resilience to risk, and transformation to a green economy. How do tradable

[7] China 2010 National Water Resources Overall Plan (2010-2030) (China State Council), 2010, http://www.sdpc.gov.cn/fzgggz/fzgh/.../P020160426386340300637.doc (accessed May 7, 2017).

water rights interact with these issues in the context of China's water law and policy reform? How can tradable water rights potentially serve as a policy instrument to facilitate China's responses to these new challenges? These questions are crucial to the reform, which need to be further explored by both researchers and policy makers.

References

Easter, K. W., & Huang, Q. (Ed.). (2014). *Water markets for the 21st century: What have we learned?* Dordrecht: Springer.

Townsend, D., & Adams, O. (2016, September). *Water rights and trading in Australia* (pp. 3–4). *Australian Environmental Law Digest.*

Wang, S. (2006). *Resource-oriented water management: Towards harmonious coexistence between man and nature* (2nd ed.). Singapore/Beijing: World Scientific Publishing/WaterPower Press.

Legal and Policy Sources

2012 China State Council Opinions on the Implementation of the Strictest Water Resources Management System, State Council Document No.3. http://www.gov.cn/zwgk/2012-02/16/content_2067664.htm. Accessed 24 Oct 2016.

China 2010 National Water Resources Overall Plan (2010–2030) (China State Council), 2010. http://www.sdpc.gov.cn/fzgggz/fzgh/.../P020160426386340300637.doc. Accessed 7 May 2017.

2016 Provisional Measures on Administration of Water Rights Trading (China Ministry of Water Resources). http://www.gov.cn/zhengce/2016-05/22/content_5075679.htm. Accessed 27 May 2017).

Index

M. Jiang, *Towards Tradable Water Rights*, Global Issues in Water Policy 18,
DOI 10.1007/978-3-319-67087-4

MIX
Papier aus verantwortungsvollen Quellen
Paper from responsible sources
FSC® C105338

If you have any concerns about our products,
you can contact us on
ProductSafety@springernature.com

In case Publisher is established outside the EU,
the EU authorized representative is:
Springer Nature Customer Service Center GmbH
Europaplatz 3, 69115 Heidelberg, Germany

Printed by Libri Plureos GmbH
in Hamburg, Germany